Supersymmetry and Beyond

Supersymmetry and Beyond

From the Higgs Boson to the New Physics

REVISED EDITION

GORDON KANE

FOREWORD BY

EDWARD WITTEN

BASIC BOOKS

A Member of the Perseus Books Group
New York

Copyright © 2013 by Gordon Kane
Published by Basic Books,
A Member of the Perseus Books Group

Books published by Basic Books are available at special discounts for bulk purchases in the United States by corporations, institutions, and other organizations. For more information, please contact the Special Markets Department at the Perseus Books Group, 2300 Chestnut Street, Suite 200, Philadelphia, PA 19103, or call (800) 810-4145, ext. 5000, or e-mail special.markets@perseusbooks.com.

Library of Congress Cataloging-in-Publication Data
Kane, G. L.
    [Supersymmetry]
    Supersymmetry and beyond : from the Higgs boson to the new physics / Gordon Kane ; foreword by Edward Witten.—Revised edition.
        pages cm
    Revision of: Supersymmetry. Cambridge, Mass. : Perseus Pub., 2000.
    Includes bibliographical references and index.
    ISBN 978-0-465-08297-1 (pbk.)—ISBN 978-0-7867-3130-5 (e-book) 1. Supersymmetry.
I. Title.

QC174.17.S9K36 2013
539.7'25—dc23
                                                                            2013006737

10 9 8 7 6 5 4 3 2 1

To Sarah, Chelsea, and Noah

Contents

Foreword by Edward Witten   xi

Preface   xv

1 Toward the Big Questions   1

To understand nature we need to know particles, forces,
and rules • Research in progress • Equations? •
Prediction, postdiction, and testing • Where are the
superpartners? • The boundaries of science have moved

2 A Little Bit About the Standard Model of Particle Physics   19

The forces: Mass, decays, quanta • The particles: Do we
know the fundamental constituents? • Particles and fields •
There are more particles: Antiparticles, neutrinos, more
quarks and leptons, Higgs boson(s) • New ideas and
remarkable predictions of the Standard Model •
Experimental foundations of the Standard Model • Spin,
fermions, and bosons • Beyond the Standard Model

3 WHY PHYSICS IS THE EASIEST SCIENCE: EFFECTIVE THEORIES   45

Organizing effective theories by distance scales •
Supersymmetry is an effective theory, too • The physics of
the Planck scale • The human scale

4 SUPERSYMMETRY AND SPARTICLES: WHAT SUPERSYMMETRY ADDS   61

Supersymmetry as a space-time symmetry: Superspace •
Hidden or "broken" supersymmetry

5 FINDING AND STUDYING SUPERSYMMETRY   81

Detectors and colliders • Recognizing superpartners •
Sparticles and their personalities, backgrounds, and signatures •
Future colliders? • Can we do the experiments we need to do?

6 WHAT IS THE UNIVERSE MADE OF?   95

What particles are there in the universe? • Is the lightest
superpartner the cold dark matter of the universe?

7 WHY IS HIGGS PHYSICS SO EXCITING AND IMPORTANT?   107

The Higgs field, mechanism, and boson • Not the Standard
Model Higgs boson

8 M/STRING THEORY!   117

What is M/string theory? • Hidden or broken or partial
supersymmetry • The role of data

9 How Much Can We Understand? 129

Testing string theory and the final theory • Practical limits? • Anthropic questions and string theory • The cosmological constant • The role of extra dimensions • The end of science?

Appendix: Predicting the Higgs Boson Mass from Compactified
M-Theory   149

Glossary   153

Some Recommended Reading   187

Index   189

Foreword

Looking back at century's end, it is stunning to think how our under-standing of physics has changed in the last hundred years. The great in-sights of the early part of the twentieth century were of course Relativity Theory and Quantum Mechanics. We learned in Einstein's Special Relativity Theory of the strange behavior of fast moving ob-jects, and in his even more surprising General Relativity we learned to reinterpret gravity in terms of the curvature of space and time caused by matter. As for Quantum Mechanics, it taught us that fact is far more wondrous than fiction in the atomic world.

Special Relativity and Quantum Mechanics were fused in Quantum Field Theory, whose most remarkable prediction—verified experimen-tally in cosmic rays around 1930—is the existence of "antimatter." Quantum Field Theory is a very difficult theory to understand even for specialists; trying to understand it has occupied the attention of many leading physicists for generations.

The last fifty years have been an amazing period of experimental dis-coveries and surprises, including "strange particles," the breaking of symmetry between left and right and between past and future, neutri-nos, quarks, and more. Drawing on this material, theoretical physicists have been able, in the Quantum Field Theory framework, to construct the Standard Model of particle physics, which puts under one roof most of what we know about fundamental physics. It describes in one framework electricity and magnetism, the weak force responsible among other things for nuclear beta decay, and the nuclear force.

Is this journey of discovery nearing an end? Or will the next half century be a period of surprises and discoveries rivaling those of the past? The questions we can ask today are as exciting as any in the past, and at least some of the answers can be found in the coming period if we stay the course.

Just in recent months, newspapers have been filled with exciting accounts of the apparent discovery of the Higgs particle at the CERN laboratory near Geneva, Switzerland. This substantially clarifies our understanding of particle physics and helps explain why two of the important elementary particle forces—electromagnetism and the weak interactions—are so different in our everyday experience. It also offers an experimentally accessible prototype of new fields that are required in modern theories of the unification of particle forces, as well as our best understanding of the early universe in the theory of cosmic "inflation." Meanwhile, ongoing experiments test the strange properties of neutrinos, and apparently show that the "little neutral particle" of Fermi does have a tiny but nonzero mass after all. Astronomers have unraveled new and challenging hints that General Relativity may need to be corrected by adding Einstein's "cosmological constant"—the energy of the vacuum. Novel and inventive dark matter searches are probing the invisible stuff of the universe. Satellite probes of fluctuations in the leftover radiation from the big bang are likely, in the next few years, to challenge our understanding of the early universe.

But one of the biggest adventures of all is the search for "supersymmetry." Supersymmetry is the framework in which theoretical physicists have sought to answer some of the questions left open by the Standard Model of particle physics. The Standard Model, for example, does not explain the particle masses. If particles had the huge masses allowed by the Standard Model, the universe would be a completely different place. There would be no stars, planets, or people, since any collection of more than a handful of elementary particles would collapse into a Black Hole. Subtle mysteries of modern physics—like space-time curvature, Black Holes, and quantum gravity—would be obvious in everyday life, except that there would be no everyday life.

Supersymmetry, if it holds in nature, is part of the quantum structure of space and time. In everyday life, we measure space and time by numbers, "It is now three o'clock, the elevation is two hundred meters above sea level," and so on. Numbers are classical concepts, known to humans since long before Quantum Mechanics was developed in the early twentieth century. The discovery of Quantum Mechanics changed our understanding of almost everything in physics, but our basic way of thinking about space and time has not yet been affected.

Showing that nature is supersymmetric would change that, by revealing a quantum dimension of space and time, not measurable by ordinary numbers. This quantum dimension would be manifested in the existence of new elementary particles, which would be produced in accelerators and whose behavior would be governed by supersymmetric laws. Experimental clues suggest that the energy required to produce the new particles is not much higher than that of present accelerators. If supersymmetry plays the role in physics that we suspect it does, then it is very likely to be discovered by the Large Hadron Collider (LHC), or its upgrades, at CERN in Geneva, Switzerland.

When Einstein introduced Special Relativity in 1905 and then General Relativity in 1915, Quantum Mechanics was still largely in the future, and Einstein assumed that space and time can be measured by ordinary numbers. Einstein's conception of space and time has been adequate for discoveries made until the present, but discovery of supersymmetry would begin a reworking of Einstein's ideas in the light of Quantum Mechanics.

Discovery of supersymmetry would be one of the real milestones in physics, made even more exciting by its close links to still more ambitious theoretical ideas. Indeed, supersymmetry is one of the basic requirements of "string theory," which is the framework in which theoretical physicists have had some success in unifying gravity with the rest of the elementary particle forces. Discovery of supersymmetry would surely give string theory an enormous boost.

The search for supersymmetry is one of the great dramas in present-day physics. Hopefully, the present book will introduce a wider audience to this ongoing drama!

EDWARD WITTEN
Princeton
November 20, 2012

Preface

If you take a little trouble, you will attain to a thorough understanding of these truths. For one thing will be illuminated by another, and eyeless night will not rob you of your road till you have looked into the heart of nature's darkest mysteries. So surely will facts throw light upon facts.

—Lucretius, *On the Nature of the Universe*
Translated by R. E. Latham, Penguin Books

Without looking at the insides of an old-fashioned cogs-and-gears watch, we have an image of what is happening inside. Few people realize that physicists now have a clear image of the cogs and gears of the subatomic universe—the stuff that makes the world run. That image is formulated in what we call the Standard Model of particle physics. It is a *description*. Someone who probes and studies a watch not only can *describe* the workings of a watch but also can say *why* the watch works—why this cog moving that one at a given ratio mimics the progress of time. Physicists are increasingly able to peer into the workings of the universe and say *why* the ingredients they study are able to create and sustain what we know as nature. The name "Standard Model" is not very elegant but it stuck as the Standard Model was proposed and tested, and now it is unlikely to change.

A nice way to learn more about a watch is to see a watchmaker disassemble and reassemble one, and a nice way to learn more about nature is to go for a walk with a naturalist. This book is meant as a walk observing the particles and their behavior that can be enjoyed by anyone with the curiosity to come along and the confidence to know that it isn't necessary to duplicate the guide's knowledge, only to enjoy it. We will stroll here not only in the known territory of the Standard Model but also along the frontier topics, where breakthroughs into even more remote regions may soon occur. For various practical and theoretical reasons, many particle physicists think that the next major discovery will be direct evidence for the property called "supersymmetry." Those reasons, and the implications if direct evidence for supersymmetry is indeed observed, are much of what the book is about.

Book One of the age-old search for understanding how the physical world works has been brought to a successful close in recent years with the development and testing of the Standard Model of particle physics. The Standard Model (summarized in Chapter 2) gives a comprehensive description of the basic particles and forces of nature, and of how all the physical phenomena we see can be described. It contains the underlying principles of the behavior of protons and neutrons, nuclei, atoms, molecules, condensed matter, flowers, stars, and more. The Standard Model has explained much that was not understood before, it has made hundreds of successful predictions including many dramatic ones, and there are no phenomena in its domain (see Chapter 3) that are not explained (though some calculations are too complicated to carry through). The last piece needed to complete the Standard Model description was the recent exciting discovery of the Higgs boson at the European CERN laboratory's Large Hadron Collider (the LHC). The particle discovered seems to have the properties needed to indeed be the Higgs boson, which physicists searched for over three decades to complete the Standard Model, though more checks have to be completed about its properties before that is finally settled. We will look at the remarkable experimental and theoretical achievements and physics involved in the Higgs boson discovery.

If the Standard Model describes the world successfully, how can there be physics beyond it, such as supersymmetry? There are two reasons. First, the Standard Model does not explain aspects of the study of the large-scale universe, cosmology. For example, the Standard Model cannot explain why the universe is made of matter and not antimatter, nor can it explain what constitutes the dark matter of the universe. Supersymmetry suggests explanations for both of these mysteries. Second, the boundaries of physics have been changing. Now scientists ask not only how the world works (which the Standard Model answers) but why it works that way (which the Standard Model cannot answer). Einstein asked "why" earlier in the twentieth century, but only in the past decade or so have the "why" questions become normal scientific research in particle physics rather than philosophical afterthoughts. So the Standard Model will be extended.

One ambitious approach to "why" is known as "string theory," which is formulated in a ten- or eleven-dimensional world. Much work on string theory has proceeded thus far by studying the theory itself rather than through the historically fruitful interplay of experiment and theory. This approach has led to significant progress—if it succeeds we will all be delighted. As Edward Witten remarks in his Foreword to this book, string theory predicts that nature should be supersymmetric. To test string theory one needs to "compactify" it so that its extra dimensions are curled up in a small space, because experiments are done in our four-dimensional world. Although one cannot walk in those extra small dimensions, their properties affect the properties of the particles and forces we do see, and there has been progress toward understanding the "why" questions, which I discuss later in the book.

This book is mainly about the physics of supersymmetry. Supersymmetry is an idea, that the equations representing the basic laws of nature don't change if certain particles referred to the equations are interchanged with one another. Just as a square on a piece of paper looks the same if you rotate it by ninety degrees, the equations that physicists have found to describe nature often do not change when certain operations are performed on them. When that happens, the equations are said to have a symmetry. Supersymmetry is such a proposed symmetry: the "super" in its name is because the associated symmetry is more surprising and more hidden from everyday view than previous symmetries. It turns out that this idea has remarkable consequences for explaining aspects of the world that the Standard Model cannot explain, particularly the Higgs physics. The most important implication may be that supersymmetry can provide a window that lets us look at string theory from the real world, so that experiments can provide guidance to help formulate string theory, and so that string theory predictions can be tested. Supersymmetry is the first chapter of Book Two.

Supersymmetry is still an idea as this book is being written in late 2012. There is considerable indirect evidence that it is a property of the laws of nature, but the confirming direct evidence is not yet in place. That is not an argument against nature being supersymmetric, because the collider facility that could confirm it (the LHC) is just beginning to cover the region where the signals could appear. I have tried to write this book so that it will remain valid and interesting if the "superpartners" (see Chapter 1) and additional Higgs bosons predicted by supersymmetry are indeed found. Of course, when we have positive signals, much can be sharpened. However, the explanations supersymmetry can provide, and the ways it can connect with string theory, and the ways we recognize and test it, hopefully will be close to what is presented here.

Some particle physics books describe several alternative approaches to learning how the Standard Model is extended. I have chosen instead to focus only on the supersymmetric direction. It is the approach favored by a variety of theoretical and indirect arguments, including the discovery of the Higgs boson itself. The issue will not be settled until superpartners are discovered, either at the LHC or in dark matter experiments. If supersymmetry is not the right answer, the experiments will tell us that as well.

This book is intended to be self-contained and largely accessible to any curious person.

I am very grateful to my most relentless editor, my wife Lois, and to Jack Kearney and Bob Zheng for thorough readings of the manuscript, and particularly to Maria Spiropulu for her extensive discussions and wisdom.

1

Toward the Big Questions

PAUL GAUGUIN TITLED WHAT HE THOUGHT would be his last painting (1897) "Where do we come from? What are we? Where are we going?" (See Figure 1.1.) The painting shows natives and aspects of life on
Tahiti, where Gauguin was then living. (It can also be viewed online
and at the Museum of Fine Arts in Boston.) In his writings he mentioned its "enormous mathematical faults" and how it was "all done
from imagination." His reflections remind us of scientists' goals in the
search for a better—and, some hope, complete—understanding of our
universe. Scientists work with mathematical constructions and imagine hypotheses while trying to grasp where we come from, what we are,
where we are going. More concretely, they try to establish why there is
a universe and how it works and why it works the way it does, what we
are made of, and how inanimate matter can give rise to conscious,
thinking people.

Every culture has asked these questions in some form and followed
some approach to provide answers. The approach that we call *science*
has led to a remarkable set of results and answers to some of these
questions, because it developed a method to study the natural world.

Figure 1.1

The scientific method originated with the Ionian Greeks more than twenty-five hundred years ago but only began to provide reliable knowledge about the world with the work of Galileo and Kepler about four hundred years ago, and with the development of experimental tests. Science makes progress by combining imagination with experimental results, by insisting on evidence.

More than one physicist, attracted by the title, has a reproduction of Gauguin's painting. But when we look at it we don't see answers that Gauguin perhaps desired, because the painting is his personal approach to those questions. Science, on the other hand, allows us to search for answers together, theoretically and experimentally, and to develop answers that become shared once they are established. I hope this book will help do that for the reader. Science poses the same questions that Gauguin and other artists ask. Its aim is to understand what lies behind the verb form "to be." Though some may perceive otherwise, this science is not the opposite of the humanities, though it is less accessible by verbal and visual images than most. Quarks can't really be represented by curly beards or white togas, electromagnetic fields can't be shown as pudgy babies with wings and bows and arrows. Equations and their solutions are the representational images of the universe's structure—the circumference of a circle, the parabola of a cannon ball are each a precise and beautiful image of part of nature. If someday we have a complete set of equations, perhaps unified into one final equation, we will have a complete mathematical image of the universe. Then we will be able to convert that to a verbal image.

Today we are at a stage where there is one main idea about the next experimentally accessible step toward understanding the basic laws that govern the universe, but it is very hard for practical reasons to get the evidence we need to learn if the idea is correct. This book focuses on that idea, called "supersymmetry." There is already indirect evidence (which will be described in later chapters) that supersymmetry is part of a correct description of nature. The recent exciting discovery of the Higgs boson is part of the evidence for supersymmetry (since, as described in Chapters 4 and 7, supersymmetry allows the derivation of the crucial properties of the Higgs boson that lead to the interactions that give mass to quarks, electrons, and W and Z bosons). I will describe the challenging and successful experimental search and the profound implications of the Higgs boson discovery. As of this writing in March 2013, it may be a little premature to explicitly refer to "the discovery of the Higgs boson," because the measurements of its properties, such as its spin, are not entirely complete. Some people prefer to say "Higgs-like boson." More data are needed, and more are being obtained. Since the current results do appear likely to lead to that interpretation, I will take the risk of writing as if the Higgs boson has indeed been discovered at the LHC.

If we understand supersymmetry and its implications correctly, direct experimental evidence for supersymmetry will be found in the next few years—possibly soon after this book is published (or, with great luck, even before). As we will see, supersymmetry is important not only as a possible previously unknown part of nature but in addition because it should allow us to probe the ultimate laws of nature much more directly.

This first chapter is meant to explain what this book is about, and to describe its assumptions and goals. It can be difficult to understand how science works, how it progresses, and how scientists working in an area become convinced that an accurate description of nature (or the universe, or the world—these words are used more or less interchangeably throughout the book) has been formulated. It can also be difficult to understand the results. The next pages are an effort to prevent misunderstandings and to lead us smoothly into our subject.

TO UNDERSTAND NATURE WE NEED TO KNOW PARTICLES, FORCES, AND RULES

In order to understand the natural world we have to know at least three things. As we probe the world we find that it consists of particles, so we have to know what the basic particles are. Over two millennia ago some Greeks correctly reasoned that the wonderful complexity of the world we see could be explained if everything was composed of a number of basic, irreducible constituents (*particles*). It wasn't until the past century that we developed the techniques needed to test ideas about the particles, and establish their existence and properties; Chapter 2 will describe the surprisingly simple results.

The particles interact to form all the structures of our world, so we have to know how they interact, what forces or interactions affect them. But even if we know the particles and forces we could not explain anything unless we also knew nature's rules, and have mathematical representations of them, to work out the behavior of the particles under the influence of the forces. For example, even if we know that two particles will attract each other and fall toward each other because of the gravitational force, we don't know how energetic a collision they will have unless we have an equation (a *rule*) to calculate their speed. Nature's rules apply for all particles and interactions. The first rules were written by Newton. Today his rules and others are integrated into two comprehensive rules: Einstein's so-called special relativity and quantum theory. These rules will not play much of a detailed or visible role in this book: you don't need to know how they work to understand the book. The important thing is that they are there in the background, as well-established and well-tested methods to calculate how particles behave when they interact through a given set of forces. They provide a framework that constrains theories and relates different parts of the theories.

One can ask whether our formulation of quantum theory and special relativity is likely to be extended or modified as progress is made toward an ultimate theory of nature. That is unlikely, at least for practical purposes. The equations and algorithms representing nature's rules are extremely well tested in a variety of situations, but the reasons to believe they will remain valid are even stronger than the explicit tests. The equations and algorithms are part of a mathematical theory that forms a coherent structure. If any part of this theory is changed, the change propagates through to other parts, and probably leads to changes in some well-tested part as well as to contradictions with some established results. (We will see later in the book that there might be some room to change the formulations for extremely high-energy interactions, though there is no reason to think such changes will occur.) To better understand the world, we also will have to understand not only what nature's rules are but why they are the rules. The effort to understand that is barely beginning to be a subject of research. But we do know enough to be confident that, for purposes of formulating and understanding and testing supersymmetry, the present knowledge of the rules is satisfactory.

As noted above, for this book the reader does not need to know much about nature's two basic rules beyond the fact that they are there, but it's worthwhile to describe them a little. The constraints of special relativity follow from two simple postulates, which can basically be stated as follows: (1) the laws of nature are the same regardless of where they are formulated and tested, and (2) the speed of light in a vacuum is the same regardless of the conditions under which it is measured. The first is obvious: it says what it seems to say, that if you work out the laws of nature on Earth, or another planet across the galaxy, or a spaceship, or anywhere, you will get the same results. The second is not so obvious. That the speed of light in a vacuum is always the same is extremely well tested by many approaches, both directly and by examination of the implications of special relativity.

The word "relativity" in this theory's name is misleading and unfortunate, since the essence of its foundations is that two things are absolute, not relative at all. The name stems from an implication of the theory, that the outcome of some experiment can be different if the experiment is carried out in laboratories that are in relative motion (e.g., one on Earth and the other in an airplane moving at constant speed overhead). However, the theory then goes on to show that when the effects of the motion are included, even for experiments in relative motion the ensuing descriptions of the results become the same.

Our main interest in special relativity here will be that it limits the form that a valid theory can take. It's a powerful constraint; for example, Newton's laws had to be reformulated because they did not originally obey the constraints of special relativity. If we need to say that a theory satisfies (or does not satisfy) the constraints of special relativity, we may use the physics jargon and say the theory is "relativistically invariant" or satisfies "relativistic invariance" constraints. Special relativity was fully formulated by Einstein in 1905. Its validity was tested both theoretically (it had to be consistent with all verified descriptions of nature) and experimentally over several decades. It is still being tested whenever new technologies become available.

The other part of the basic rules, quantum theory, was formulated by several people during the years 1913–1927. For this book we do not need to know much about how quantum theory works, only that it is there and tells us how to calculate the behavior of particles if we know the forces that affect them. Later in the book we'll look at a few properties of quantum theory that we need for specific purposes. Special relativity and quantum theory have been successfully combined into a "relativistic quantum theory." Whenever I use that phrase it is only meant to tell us that the ideas under discussion have been successfully formulated to simultaneously obey the rules represented by special relativity and quantum theory.

Before about 1965 we knew very little about what basic particles were the constituents of matter. We did know what forces existed, but not how they worked to shape the world. By the end of the 1980s we had learned what the basic constituents of matter are, and understood how particles and forces function to make our world. That body of knowledge is called the "Standard Model" of particle physics; it is the subject of the next chapter. It provides a well-tested description of how our world works. Normally I'll refer to this as the Standard Model, leaving off "of particle physics." If nature is supersymmetric, the Standard Model will be *extended* to become the Supersymmetric Standard Model; the Standard Model will not be wrong but, rather, will become a part of a more complete description of nature. That is the normal way science progresses. Once an area is well tested and established, it is not put aside as the description of nature broadens but extended and integrated into the new picture.

The Standard Model is not a model in any conventional sense of that word, but the most complete mathematical theory ever developed. For physicists, "theory" does not mean what it might in everyday usage. Most generally, a theory is a principle or set of principles that imply properties of the natural world. In physics a theory typically takes the form of a well-defined set of equations that relate some symbols. The symbols represent parts of the natural world. The equations can be solved to learn the behavior of the quantities, and thus the predicted behavior of the parts of the world represented by the quantities. For example, consider Einstein's famous equation $E = mc^2$, a consequence of special relativity. Here the symbols are E, m, and c. E represents an amount of energy, m a mass, and c the speed of light. If an amount of mass m is converted into energy, this equation tells us how much energy is obtained. Solving this equation for E is easy: we just multiply m by c^2. In general, solving equations can be much harder. As an aside, $E = mc^2$ is one of the many reasons we are confident that special relativity theory is correct, because it is tested daily at colliders and in nuclear reactors.

In science the use of the word "theory" carries a loose implication that the predictions are largely tested and verified. When scientists start to study an area of the world they first make models to guide thinking, suggest experiments that might be relevant to making progress, and allow quantitative predictions. Models usually begin as limited mathematical descriptions of how some aspects of the world behave. If they work well, more phenomena are added. Later one improved model turns out to describe nature rather well, and is often named the "standard model." Even later a version of that model is so robust and well tested that it becomes the theory of that area. But it is already called the "standard model" so the name stays, even though it is really not a model any more. I take account of that development in a limp way by capitalizing "Standard Model" out of respect for the many successes now of the Standard Model. In everyday usage "theory" means something very different, a kind of vague idea about how to explain something. We might say "one theory about the falling crime rate. . . . " The use of "model" in science and in everyday language is similar, at least until a model is found that successfully describes nature, but the use of "theory" is very different, and can cause confusion. In this book we will stick to the scientific use only. The correspondence between any theory and its equations and the structure of the physical system it represents is so exact that physicists conventionally (perhaps confusingly) regard them as interchangeable in writing and discussions.

RESEARCH IN PROGRESS

It is important to distinguish carefully between those areas of science that are conceptually and experimentally well established and those that are still developing. As science develops, it becomes possible to study some previously inaccessible part of the natural world. Often a fresh area opens because of technological innovations that become available: before the microscope was invented, it was not possible to study phenomena smaller than what we could see with bare eyes. Sometimes the opportunity to study new areas depends on the past successes of science itself, since subfields build on earlier ones and, sometimes, on the introduction of new concepts. Eventually the foundations of a description of that aspect of the natural world are worked out and experimentally verified. That has occurred many times—for example, with optics, electromagnetism, thermodynamics, atomic physics, the Standard Model of particle physics, and other areas.

As a subfield is being studied and worked out, models and explanations and proposed ideas may change. Sometimes experiments don't work correctly the first time they are done, but experiments can be and are constantly redone and improved, so questions can be settled by improved experiments rather than by argumentation. Many of today's experiments are commissioned with very detailed simulations and calibrations done ahead of time, and perform very well immediately. The detectors at the CERN Large Hadron Collider (LHC, described in detail later) worked superbly from the beginning. Progress in computation and simulations has changed the experimental world.

In all experiments, the variables that the equations depend on take values in some specific range—variables such as velocity, astronomical distance, resolution of a microscope, and so on. Historically, once a theory is successfully tested in a given range of the variables it depends on, it will always work in that domain of the variables. However, if the theory is extended to new values of the variables—faster speeds, smaller or larger distances, and so on, it may work successfully in the new domain, or it may not. When it does not, eventually a new theory will be found that works in the larger domain. That new theory will "reduce to" the older one when the values of the variables are restricted to the older range. The older theory is not "wrong"; rather, it is extended to become a part of the more general one. Eventually the entire range of variables may be covered; for example, the description of motion has now been tested for all possible speeds, from zero to the speed of light, and it will not change in the future. All the well-known examples of theories or rules once thought to be valid and later found to be inadequate fit this general picture.

Once upon a time, a few hundred years ago, there was no method called science. We have had to learn how to do science at the same time we have learned the results of science. The process just described was not originally understood. For example, Newtonian science was extremely successful, and people naturally assumed it held for the whole universe. Only later did they realize that it was only approximately valid, holding when velocities and masses were not too large, and also failing for systems of atomic size. By the 1930s, physicists had finally realized that every theory had to be tested again whenever it was extrapolated beyond the range of variables where it was known to be valid. On the other hand, the need for some changes led some philosophers and historians to claim that the results of science will always change. That is wrong. For example, heat is due to the motion of molecules and always will be. The energy levels of isolated atoms will never be different, that we are made of atoms and that protons are made of quarks will not change. We could make a very long list.

The fact that the foundations of many areas of physics (and other subfields of science) are in place does not mean they are no longer active areas of research. On the contrary, many people become excited about working out the implications of the new foundations. Having the foundations in place means that the basic equations that govern behavior in that subfield are known. But as I have already discussed briefly and will consider again, finding the solutions of the equations can be interesting and difficult. Having the solutions for one system does not guarantee that one will have them for the next one, so most subfields continue to have interesting problems to solve long after their basic principles are understood. For complex systems, new predictions and results can emerge from further study or experimentation even after a long time. To put it differently, most areas generate many applications once their basic principles are understood, and will continue to do so.

For our purposes, the main distinction to keep firmly in mind is that between subfields for which the foundations are in place and subfields for which research is in progress. We expect supersymmetry and string theory to describe nature, but it could turn out that this is simply not how nature behaves. That is very different from the Standard Model, which is now known to describe how nature does behave in its domain. The Standard Model will be extended to become part of a larger theory, but in its domain it will not change.

Today, graduate students in physics are taught more about relativity than Einstein knew. Thousands of scientists use quantum theory daily. Once an area is mature, ways to explain it are developed. Writing or talking about changing areas is challenging mainly because the final answers are not yet in place, or at least not confirmed and tested, but also because there has not been time or inclination to develop nontechnical analogies and explanations. When (if) supersymmetry and the final theory that it moves us toward are in place, they will be easier to explain than they are now. Part of the general problem in communicating scientific results to the general public is that the newsworthy results are the new ones. The results may be modified later as experiments improve and ideas continue to be tested. It is worth stressing the self-correcting and tentative nature of the research that exists until a subfield is understood, and it is worth stressing that subfields eventually stop changing.

For the nonscientist, the most important and interesting aspects of any area where research is in progress are not the details but what questions are under study and what kinds of answers are being considered. Subjects that previously were the domain of philosophy have become accessible to science as new techniques were invented and more of the world was understood. The current favored answers may change (or may not). Historically, the questions were eventually answered once a subfield became a scientific research area. If history is a useful guide, that typically takes a few decades. Some areas are so speculative, so likely to survive only in dramatically modified forms, that articles and books about them may confuse general readers more than they inform them. Supersymmetry is now a sufficiently mature area, and sufficiently close to confirmation if it is indeed a part of the correct description of nature, that a wider understanding of its content and implications is both possible and worthwhile.

EQUATIONS?

Sometimes I teach a science course suitable for any students who are curious about what we have learned about our universe. It covers physics from before the big bang to the origin of life, including a considerable amount of the history of science and the impacts of science on civilization and vice versa. I always make a bargain with the students that they seem to like. It is all right to occasionally use equations to illustrate some subject, and even to use high school math to carry out derivations. But I won't test them on it. Under those conditions they aren't afraid of the mathematics, and they pay attention to the equations and the derivations and learn more. One of the remarkable things about derivations is that sometimes you start from an input you understand and agree is established, and follow a few correct steps, and end up with a surprising result that you would never have believed if you followed only a verbal argument. I hope the reader, too, will agree to not be put off by the few simple equations in this book. If the explanations are not clear enough to get you through them it's my fault, and you won't be tested on them. The intent is to help you understand the results and to explain why we accept them.

There is a *major* distinction between the properties of an equation and its solutions. We think of a theory as an equation (or several equations). On the other hand, our world is described by the *solutions* of the equation. That's how it always is in science. The principles are embodied in equations. The actual aspects of the world are described by solving the equations and finding out how the solutions behave. It can easily happen that the equations have some properties that a given solution does not have. Since that is unfamiliar to most readers, let's explain it with an example that does illustrate this idea, though it does not correspond to any real situation. However, as is common with examples and models in science, let's use one that has similarities to a real puzzle, why the world contains three particles that apparently are the same except for their masses. Even though the equations of the example aren't realistic, they illustrate how the true ones might behave.

Suppose a theory tells us that the equation relating the masses of the three particles (call them electron and muon and tau) is this:

$$EMT = 64$$

Here E stands for the mass of the electron, M for the mass of the muon, and T for the mass of the tau. The equation just says multiply the three masses and you get 64. (We're ignoring units here.) This equation is totally symmetric: if you exchange any of the masses, you get the same equation back. Having the laws be symmetric is generally very desirable. If you didn't think about it, you might guess that the individual values of the masses would then all be equal since they come from solving a symmetric equation.

Let's list solutions of this equation. For simplicity consider only solutions for which E and M and T are positive integers. There are several sets of three numbers whose product is 64. One solution has $E = M = T = 4$, since $4 \times 4 \times 4 = 64$. For this solution all the masses are indeed equal, so it is a symmetric solution. But another solution has $E = 1$, $M = 2$, $T = 32$, which again multiply to 64. There are lots more—for example, (E,M,T) can be $(1,1,64)$; $(1,4,16)$; $(1,8,8)$; $(2,2,16)$; $(2,4,8)$. In each case just multiply and you get 64. (In our example, E, M, and T represent particles that are the same except for their masses, so we define E to be the lightest and T the heaviest.) If all of them are solutions, how do we know which solution should actually describe nature? In this "toy" case we can solve the equations to find all the solutions, but in the real world solving the equations is often extremely difficult. And solving the equations is not enough, because we want to know how nature ended up being described by one of the solutions and not the others.

To show the possible power of data, suppose that we knew from experiments that none of the three masses were equal to one another; then, only three of the solutions could be correct. If in addition we knew from the data that none of the masses was more than about five times heavier than the others, we would be led to a unique solution: $(2,4,8)$.

From this instructive little example we have seen several things. The principles and laws are embodied in equations that often are highly symmetric (e.g., supersymmetric). The world (i.e., the particles and how they interact) is described by the solutions to those equations. The solutions do not need to show the symmetry of the equations—in general, they do not. The symmetry of the basic law is hidden if we can observe only the solutions. Some of the most difficult challenges scientists face in going from the world we observe to learning the laws that govern it arise because the laws have hidden symmetries that are not apparent in the world (e.g., the particle-interchange symmetry of the Standard Model described in Chapter 2). As we will see, supersymmetry itself is thought to be a well-hidden symmetry. Thus even if we have the basic laws, it is unlikely that we can figure out which solution describes the world unless we have some relevant data from experiment and observation.

To be fair, though, it could happen that the theory is so powerful that it successfully picks out the correct solution at least in principle. Let's extend our example. Suppose the theory produces a second equation, $E + M + T = 14$, in addition to $EMT = 64$. Both equations are entirely symmetric, but the only solution that simultaneously satisfies both equations is the one we found above, $E = 2$, $M = 4$, $T = 8$, and we did not need any data to learn that. Note that the two fully symmetric equations have one common unique solution and it does not show a symmetry—perhaps our world is like that. We still need data to test whether any solution we find is the actual one that describes nature.

PREDICTION, POSTDICTION, AND TESTING

The goal of science is to achieve understanding. One method or tool to move toward understanding is to make predictions and test them. Ideas and theories have testable implications about the world. Tests of predictions can have three kinds of outcomes. If the prediction is verified, our confidence in the ideas that led to the prediction is strengthened. If the prediction is wrong, the theory must be modified or discarded. But frequently the situation is a more subtle one. Often a prediction can be made in principle but depends on some quantities (e.g., masses) that are incompletely known. Then an iterative process occurs. Encouraging results entice more people into measuring relevant quantities or calculating needed ones more accurately. The prediction and its tests are sharpened over time. Eventually the prediction is tested.

Another subtlety is that it can be meaningful to "predict" something that is already known, such as the mass of the electron, or the existence of the force of gravity, or the fact that we live in three space dimensions. Such predictions are sometimes called "postdictions." They are meaningful because they occur in a theory that uniquely requires such an outcome. Such results can be powerful tests of the theory even if they were already known before the theory was formulated. In other cases, a theory can address an issue such as the mass of the electron (while earlier theories could not even in principle answer the question) but be unable to provide a definite numerical answer or prediction because some input information is not yet known. In this case, it may be possible to predict that the outcome is in a certain range that includes the known result. That outcome is very encouraging compared to predicting the wrong range, or not being able to address the issue at all. Sometimes in this sort of situation, instead of saying that the theory predicts the result we say that the theory is consistent with the result. Eventually better understanding of the theory and newly available input information lead to good predictions and tests.

There is another reason that predictions and tests are less clear than naively might be hoped. All measurements have experimental errors. Errors have statistical uncertainties: if you flip a coin six times, occasionally there will be three heads and three tails, but sometimes four and two, and occasionally five and one, and rarely six heads or six tails. All real measurements can fluctuate similarly. There are also what is called "systematic errors." These happen when the apparatus is not perfectly understood. In principle, systematic errors can be understood and avoided, but often it takes time until such sources of errors are accounted for. Most predictions for experimental tests also make some assumptions, particularly tests of new ideas. Eventually most assumptions can be independently tested or avoided, but usually not for the initial tests of newer ideas. It would be nice if all tests were clear and conclusive, and eventually in physics they are. But the more usual situation is the cloudier sort described in this section.

WHERE ARE THE SUPERPARTNERS?

The particles of the Standard Model are the electrons and quarks (explained in the next chapter) that we and all the objects in our world are made of. We'll see later that the main test of the validity of the idea that our world is supersymmetric is the existence of a set of previously unknown particles, called superpartners. At the time of this writing, in 2012, the superpartners have not yet been directly observed. Where are they? Why don't we see them? We think there are two parts to the answer. All but one of the superpartners are expected to be typically at least as heavy as the heavier of the Standard Model particles and, therefore, like them, to decay rapidly into lighter particles. ("Decay" is explained in Chapter 2.) If nature is indeed supersymmetric, the superpartners can be created in collisions at laboratories, as discussed in detail in Chapter 5, and the lighter ones are also created about once every few minutes in collisions of cosmic rays at the top of the earth's atmosphere somewhere around the world. But they decay rapidly and there is no way to detect them in the atmosphere. Colliders and detectors at laboratories are now achieving the energies and luminosities (amounts of data) and sensitivities needed to explicitly detect the superpartners, at least if our thinking about their properties is more or less right. They might have been lighter if we were lucky, so they could have been observed at earlier colliders, but that did not happen.

We think that when superpartners decay there has to be a superpartner among the particles they decay into. They have to decay into lighter particles since their mass turns into the combined mass and energy of the decay products. So eventually each superpartner decays into Standard Model particles such as electrons or quarks or photons, plus the lightest superpartner (since there has to be a superpartner). The lightest superpartner is expected to be stable because there is no lighter one into which it can decay (though this is not guaranteed). Then all of the lightest superpartners that have been created during the big bang, and in collisions, and in decays of heavier superpartners since the big bang, should still be around and spread throughout the universe (except for a calculable small number that can annihilate with each other). This is the subject of Chapter 7. The estimates are that these relic lightest superpartners can make up part or most of all the matter in the universe, the dark matter, with about one in every grapefruit-sized region around us. That may be where the superpartners are. There are indirect ways in which we can observe them. In addition, they can be produced at the LHC.

THE BOUNDARIES OF SCIENCE HAVE MOVED

One of the innovations in thinking that made modern science possible was focusing on how the world worked, rather than on why the world was the way it was. Four centuries ago, "why" was left in the realms of religion and philosophy. "Why" questions were recaptured by science initially in biology, when Darwin was the first person to make them scientifically legitimate, two and a half centuries after modern science began. It took over a century longer for physics to begin dealing with the "why" questions.

Before about the 1980s, the questions that physics could address clearly had limits. Big questions such as where the laws of nature came from, and why there was a universe at all, were out of bounds. People could argue that each question answered would give rise to more questions, so there would always be voids in scientific knowledge; that science would always have to take some fundamental aspects on faith; that some things were unknowable. What has changed is that now all of these big questions have become technical research questions. We don't know yet whether they will have scientific answers, but they are finally research questions. Most of the people working on them expect answers. Today a number of active researchers don't expect that there is anything necessarily unknowable concerning fundamental questions about the physical universe, nor that new questions about the ultimate laws of nature will necessarily continue to arise. We will return to these issues in Chapter 9. Supersymmetry does not itself provide the final answers to these big questions; but if our current ideas are right, it will be crucial to provide a way to ensure that these ultimate questions can be studied as normal science, and that proposed answers can be tested in normal ways. The purpose of this book is to help the reader understand how we will find out if nature is indeed supersymmetric, and also how the big questions can be studied if it is.

2

. . . .

A Little Bit About the
Standard Model of Particle Physics

To better understand the reasons we think supersymmetry will be discovered soon and, if so, how it extends our present description of nature, we need to know a little about the main achievements of more than two centuries of physics. Past study led to the establishment of the Standard Model of particle physics, a complete description of the basic particles and forces that shape our world. The world we see is actually built of three particles, the electron and two particles similar to the electron called quarks—a description simpler than any previously proposed throughout history. The quarks bind into neutrons and protons, which bind into nuclei, which join with electrons to make atoms. There are some additional particles—antiparticles, neutrinos, more quarks and more particles like the electron, and Higgs bosons. We understand why some of the additional particles exist, but not others. We don't need to know much about them for our purposes. Next we consider a property of all particles called "spin," and how it leads to categorizing particles into two groups, called "bosons" and "fermions." We need to pay attention to this property since supersymmetry relates them. Finally we look at reasons why the Standard Model is not expected to be the final stage of particle physics even though it successfully describes phenomena and experiments. Of course, we could just say supersymmetry is a cool idea and we should search for it in nature, but the search requires major facilities and great effort so it is good to examine the motivations for supersymmetry. The Standard Model is an awesome theory. Developing it and testing it have been great achievements of many physicists. This chapter describes in part the Standard Model and its successes.

When Gauguin painted "Where do we come from? What are we? Where are we going?" we knew only the electrical, magnetic, and gravitational forces. There was controversy about whether atoms existed. The first particle to be discovered, the electron, had just been found. Radioactive decays of nuclei ("radioactivity") had just been noticed. These decays could not be explained by the known interactions, so physicists realized that another force was needed to describe nature's behavior. It was called the "weak" force, because its effects were rare and essentially never occurred when two interacting objects were separated by distances larger than an atom. In 1911 atoms were discovered to have a nucleus, and heavier nuclei were found to have a number of protons in them. Knowing that the repulsive electrical force pushed the protons apart, physicists deduced that yet another force, the nuclear force, must exist to bind the protons and neutrons together into a stable nucleus. The effects of the nuclear force also can be felt only at tiny distances, no larger than a nucleus. Although Newton had correctly foreseen three centuries ago that there could be forces that had effects only at small distances, finding evidence for them took over two centuries.

These five forces (electrical, magnetic, gravitational, weak, nuclear) account for all we observe in nature. The interactions of particles with a "Higgs field" (more about this later, especially in Chapter 7) can be thought of as giving mass to the particles, so one can think of the Higgs interaction as an additional force. There may in a sense be other forces, but they are not relevant for the behavior of particles or for how particles combine to make up the world around us. (For completeness, let me mention two at this point. One is a force that recent evidence suggests causes the space-time of the universe to expand more rapidly than it would if only the gravitational force affected the expansion. This force does not affect individual particle behavior at all. For historical reasons it is called the "cosmological constant." Others are possible forces that have effects only at extremely tiny distances, far smaller than the size of a proton.) We will briefly return to possible additional forces later; the five known forces are the important ones for our purposes. Any others do not affect how our world works, though understanding them may be essential to achieve a complete picture of the laws of nature.

Once Charles-Augustin de Coulomb showed, over two hundred years ago, that the electrical and gravitational forces depend in the same way on the distance between interacting objects, and therefore that the formulas describing them have the same form, physicists have tried to unify our understanding of these forces. In the second half of the nineteenth century, using the work of Michael Faraday and others, James Clerk Maxwell succeeded in relating the electrical and magnetic forces, in the sense that electrical forces that vary in space or time generate magnetic ones, and vice versa. By the 1960s we knew of the five forces, with the electrical and magnetic ones unified into electromagnetism. There was no mathematical theory at all of the weak and nuclear forces. A few regularities of their behavior were known. By the 1980s the Standard Model had emerged and been well tested. It was a complete description of the weak, electromagnetic, and strong forces, fully consistent with quantum theory and special relativity. The progress over two decades was spectacular: in that short period we went from a crude awareness of the weak and strong forces to their comprehensive description.

The picture of the electromagnetic force that emerges is that electrons, and any particles that have electric charge, interact by exchanging photons: the photons can carry energy between the electrons; two electrons can scatter off one another by exchanging a photon; and an electron and a proton bind by exchanging many photons, which provide an attractive force that keeps the electron and proton connected in a stable object, a hydrogen atom. All the forces work in a similar way. The gravitational force arises from the exchange of gravitons. The analogous particles for the weak interactions are called W and Z bosons, and for the strong force they are called gluons. In all these cases, we speak of the photon, W and Z bosons, and gravitons as *mediating* the forces. The photons and gravitons are quanta of the electromagnetic and gravitational fields. The W and Z and gluons are also quanta, of the less familiar electroweak and strong fields. There is one more subtlety: the name "strong" force is used here. Because of the history, I cheated a little when I listed the nuclear force above. It turns out that the nuclear force between protons and neutrons that binds them into nuclei is not the fundamental force. Rather, the basic force is the strong force between quarks, and the nuclear force is a kind of residual effect after quarks are bound into protons and neutrons. I'll return to the strong force after I describe the quarks.

THE FORCES: MASS, DECAYS, QUANTA

Two frequently used words about particles that can be confusing are "mass" and "decays." For our purposes, "mass" essentially means weight, and we don't need a more precise definition. Some particles—for example, photons—have energy but no mass. The masses of particles are measured mostly by bouncing particles off of each other, since how much they bounce is related to how heavy they are.

Almost all particles are unstable and decay into others. The word "decay" has a technical meaning in physics: one particle disappears, turning into typically two or three others. A major difference in the way "decay" is used in physics (compared to everyday life or biology) is that the particles in the final state are not already in the decaying particle in any sense. The initial particle really disappears, and the final particles appear. The photons that make up the light we see provide an example: the photons that come from a light bulb when it is turned on are not particles that were in the bulb just waiting to come out, and photons that enter our eyes after bouncing off an object are absorbed by the molecules in our eyes and disappear. All particles can be created or absorbed in interactions with other particles. The Higgs boson was detected mainly by disappearing into two photons, seen in the detectors, whose energies combine into the mass of the Higgs boson. Almost all particles were created during the big bang, and particles can be created in collisions (e.g., at colliding beam accelerators). Which particles can appear or disappear in any interaction is not arbitrary but, rather, fully determined by the interactions described by the Standard Model (and its extensions such as supersymmetry).

Two constraints keep a few particles from decaying. First, the total amount of energy in any process must not change—we say energy is conserved. Imagine a decaying particle at rest—its energy is just its mass. That mass is divided into the sum of masses of the particles created in its decay, plus some energy of motion of those particles. Since energy can't appear from nothing, the particles created in the decay must be lighter than the decaying particle. Second, most particles carry some charge (electric charge and some similar charges associated with other forces). Charges are also conserved—the total amount of charge can't change in a process. Because of these two constraints, several of the lighter particles—for example, electrons—do not decay. Whenever basic particles are able to decay the decay is typically rapid, occurring in a tiny fraction of a second.

Electromagnetic waves carry energy from antennas to our radios, or from light bulbs to our eyes. One of the things quantum theory has taught us is that the energy is carried in little chunks (or quanta), made up of one or more photons. Any electric charge sets up an electric field around itself, and when that charge oscillates (say, in an antenna), it radiates the electromagnetic wave or the photons. We have also learned that there is a gravitational field associated with mass, and there are other fields associated with matter, with electrons and other particles. All the particles can be thought of as the quanta of the fields, quarks and electrons as well as photons and gravitons. This book won't make any technical use of such ideas, but every time I talk of some kind of field I will associate quanta and therefore particles with it, and vice versa.

THE PARTICLES: DO WE KNOW THE FUNDAMENTAL CONSTITUENTS?

When viewed from the particle side, the Standard Model is remarkable. What are we made of? What happens if we keep cutting an apple, or a person, into smaller and smaller pieces—do we reach a smallest piece? What could it be? What are the stars made of? Everyone has wondered about such questions. The answer is: Everything we see in the universe, from the smallest cell to flowers to people to stars, is made of three kinds of matter particles, bound together by gluons and photons. When we get to big objects such as planets and stars, gravity binds too. The three matter particles are the familiar electron, and two particles similar to the electron, called quarks: the up quark and the down quark. "Up" and "down" do have a technical meaning, but it's not important here. All of the basic particles carry various amounts of charges: electrical charge and weak charge and strong charge. The weak and strong charges are somewhat like electrical charge, but they have no effects outside atoms so they are not familiar to us in everyday life. The main difference between electrons and quarks is that the quarks carry the strong charge, so they interact via exchanging gluons, whereas electrons do not interact with gluons at all. Electrons and quarks all have electrical charges and masses (weights), in different amounts. Sometimes electrons are denoted by "e," up quarks by "u," and down quarks by "d."

How do these seeds form our world? The quarks bind together to make neutrons and protons. Neutrons consist of an up quark and two down quarks, while protons have two up quarks and one down quark. The quarks interact via the strong force, mediated by gluon exchange. The edges of protons and neutrons are not sharp: gluons range outward a little before they are brought back by the attractive force of the quarks. Gluons that are a little outside the proton and neutron in turn provide an attractive force felt by other protons and neutrons, which is the nuclear force that binds the protons and neutrons into nuclei. The nuclear force is strong enough to hold together many nuclei, with one to ninety-two protons and varying numbers of neutrons. The electrical repulsion felt by one proton (due to all the other protons in the nucleus) increases as the number of protons increases; with more than ninety-two protons, the nucleus becomes unstable. That is why there are only ninety-two naturally occurring chemical elements. The electromagnetic force, mediated by photons, binds electrons to nuclei to make atoms, the atoms of the chemical elements. Outside an atom there is a residual electromagnetic force (analogous to the residual strong force outside protons and neutrons described earlier in this paragraph) that binds atoms into molecules. Atoms and molecules build up rocks, cells, and all of the world around us. All of the marvelous complexity and color and structure of our world arises from these simple foundations. This description, based on a century of experiments and theory, is not just qualitative or an analogy; it is a complete quantitative mathematical theory.

Always in the past when objects were studied at smaller distances they turned out to have structure. The atoms of the ninety-two chemical elements were not the atoms invented by the Greeks as the basic indestructible units of matter, since those turned out to have a nucleus surrounded by electrons. The nucleus turned out to be made of protons and neutrons. Protons and neutrons turned out to be made of quarks bound by gluons. Why do we think that the electrons and quarks are the true Greek "atoms," and that despite a history of smaller and smaller units the progression stops with them? There are three kinds of reasons that lead most particle theorists to think we have finally reached the end of the line.

The first, significant but less compelling than the others, is that experiments have looked at whether electrons, quarks, W bosons, and gluons show any evidence of structure by many methods, and not found any. They have probed perhaps ten thousand times further than it took to see structure in the past, but electrons and quarks continue to behave as point-like objects with no parts. This first reason is less compelling because experiments can probe only as deeply as the most energetic beams allow, while the next reasons use the consistency of relativistic quantum field theory to extrapolate further.

The second reason is that always before the theory of atoms or nuclei or protons did not agree with experiments or even make sense unless these objects had structure at smaller distances, but the Standard Model is different. Because the Standard Model is a quantum theory it is possible to ask how the forces behave at smaller and smaller distances, and to do calculations to answer the question. Suppose that the basic particles were indeed point-like. How strong would the forces be if we could examine them not only down to almost 10^{-18} meters where experiments can presently be done, but at a million million times smaller distances? The calculations show that in the Standard Model the interactions that bind the quarks become weaker at smaller distances. The Standard Model does not lead to inconsistencies with the data or the theory itself, no matter how small the distance that is probed. The Standard Model theory does not even have any possible way to include structure at smaller distances, nor parameters for the size or possible parts—it is a consistent theory without such parameters. The first reason, while very suggestive, cannot be compelling because the next experiment may show evidence of structure (though it's not expected to). But having a theory that is confirmed by many experimental tests and can be extrapolated to much smaller distances is a very powerful and convincing argument.

The third reason is also very compelling, and we will consider it in more detail later in another context, but it is worth mentioning here. As stated above, we can calculate a prediction for the behavior of all the forces at smaller and smaller distances even though we cannot do experiments in the relevant regimes. What is found is that *if* the electron and quarks are structureless, then the different forces become more and more similar in strength at smaller distances, and at a sufficiently small distance they become indistinguishable, thus suggesting how the two-century-old goal of unifying the forces might be realized.

If electrons and quarks have structure, this does not happen. Unless this unification is a coincidence, it is a strong argument for no further structure.

When I say that the quarks, electrons, photons, and other particles that mediate forces are the true fundamental particles, I do not mean they will not be reinterpreted—for example, as vibrating strings or related objects instead of point-like objects. Perhaps the particles will be thought of as analogous to the musical notes made by a vibrating violin string, with the string as the fundamental thing. But there will still be a vibration identified as an electron, another as an up quark, another as a photon, and so on. These basic constituents are not expected to be replaced by others. We will follow the Roman poet Lucretius, who wrote so presciently about science and nature two millennia ago, to " . . . reveal those atoms from which nature creates all things . . . [and] call them the 'final particles.'"

PARTICLES AND FIELDS

Newton's theory of gravity worked wonderfully to describe the moon and projectiles and the tides and much more, but it had a conceptual flaw. It seemed to work by some kind of instantaneous "action at a distance." How did the moon know at some instant to turn its path and stay near the earth instead of going straight off into space? Many people, especially philosophers, were upset about that and resisted accepting the gravity theory. Physicists are used to thinking about how something works separately from why it works that way, so they went ahead and developed and applied Newton's theory. It took over two centuries to fully understand what was actually happening.

The initial development wasn't intended to solve the problem but turned out to be the first of several major steps that did solve it. In order to simplify solar system calculations, where the effects of all the planets on each other had to be included, people developed a procedure different from Newton's but equivalent. Instead of adding up the forces on a planet from all the other planets and the sun, which was very complicated to actually do, they considered one planet at a time as if it were the only object in the universe, and imagined that it set up a gravitational "field" through all space, defined such that if another planet were put at a given point the resulting force between the two would be exactly the Newtonian one. At a given point in space one could add the fields from different bodies and obtain the total field, which was then what would be felt by another planet if one were there. Originally this was viewed as just a trick to simplify calculations.

It was Michael Faraday who convinced physicists to think of fields as real physical things rather than as calculational devices. He did that in the context of electrical and magnetic fields. Most of us have seen what happens when iron filings are put on a piece of paper over a magnet: they align themselves into a visual representation of the magnetic field. We now think of electrical and magnetic and gravitational fields as being as real as rocks and people. Normally we are not aware of those fields, but you can confirm that they are all around you by turning on a radio.

The next stage came when Einstein demonstrated that no signal or bit of information could move faster than the speed of light, and in particular that electromagnetic and gravitational fields spread out at exactly the speed of light. That's very fast, but not instantaneous. For classical (pre-quantum) physics, the action-at-a-distance problem was solved. The final stage came in the context of quantum theory. First photons were interpreted as the quanta of the electromagnetic fields. Then it was realized that electrons and all particles should be viewed as indivisible quanta, each of their own fields. The effects of the fields are transmitted by the quanta. Today we think of the fields as the basic ingredients of the universe, and often talk of the particles and the fields interchangeably. For clarity we remark that the phrase "quantum jump" that has entered everyday vocabulary is what happens when an atom or any system undergoes a transition from one stationary energy level to another by emitting or absorbing a photon or other quantum.

The modern way to describe nature at the level of the particles is for theorists to write down an expression called a Lagrangian (named after

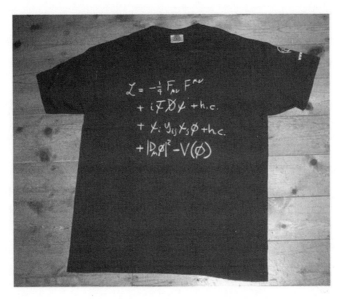

Figure 2.1 This shows the Lagrangian of the Standard Model. From it students after several courses can deduce the existence and properties of protons, atoms, molecules, and the world that we see. (The gravitational force has to be added to include planets, stars, etc.) Credit: CERN.

Joseph Louis Lagrange, who first formulated Newton's theory in this way). The Lagrangian contains the symbols representing the quantum fields for all fundamental objects—electrons and quarks, photons and gluons, and W and Z bosons—and only for the fundamental objects. All composite objects, anything with structure, such as protons, nuclei, atoms, and so on, are built up from the particles described by the Lagrangian. Given the Lagrangian, the rules of quantum theory tell us how to calculate the behavior of all systems in principle, though in practice some are very complicated. Knowing the Lagrangian means knowing the fundamental particles and forces. The achievement of the Standard Model was to learn the Lagrangian of the world that we see. In this book I will use only a little of the jargon of the Standard Model. It is helpful to keep in mind that all the particles and forces of the Standard Model are included in one formula from which all the predictions for the particles behavior can be calculated. We had to give this formula a name, "Lagrangian." It fits easily on a T-shirt (see Figure 2.1), available in the CERN store.

THERE ARE MORE PARTICLES: ANTIPARTICLES, NEUTRINOS, MORE QUARKS AND LEPTONS, HIGGS BOSON(S)

While it is true that everything we *see* is made of the electron and up and down quarks, research has uncovered the existence of some more particles. We do not see them, and they do not directly show up in our world, but we can do experiments to uncover them. We understand why some exist, but others (most of them) are a mystery. For completeness, here is a brief list of some additional particles: antiparticles, neutrinos, more quarks and leptons, and Higgs bosons.

Antiparticles

In 1928, Paul Dirac combined quantum theory and Einstein's special relativity in the description of the electromagnetic interaction. In the process he showed that the resulting equations had a surprising property. We have seen that particles are represented by the solutions of the basic equations. Dirac showed that if the behavior of one particle—say, an electron—was described as a solution of the basic equations, then the equations must also have another solution with the same mass as that particle, but with all charges opposite in sign. In the case of the electron, with a negative electric charge, the new particle would have a positive electric charge. If one solution exists in the real world, so must the other. It is actually similar to saying that the square root of 4 is both +2 and −2, since either of them squared gives 4. To make an equation we could write $x^2 = 4$, with solutions $x = +2$ and $x = -2$. If one is a solution, so is the other. Similarly, if one particle exists, so must the other. The two particles are called antiparticles of each other. Dirac's result holds for all quanta or particles. If a particle has no charges, such as a photon, it is its own antiparticle: it and its antiparticle are not distinguishable.

This was the first major example in which a previously unknown particle was predicted by theoretical arguments—not just one particle but a doubling of the number of particles, an antiparticle for each particle. What is amazing is not that "antimatter" exists but that a purely theoretical argument predicted the existence of unknown particles and later they were actually found. Human thinking as well as guidance by a theory were required to find a previously hidden part of nature. We will see that supersymmetry predicts another doubling of the number of particles.

There is a certain popular mystique about antimatter that is not justified. The particles and their antiparticles are all just particles with different charges. The antiparticles have all been observed. Which is called the particle and which is called the antiparticle is just a convention. Antiparticles will individually be as permanent as the associated particles, and behave the same way the particles do (once the opposite sign of the charge is taken into account). For example, antielectrons (often called positrons), and antiprotons can bind to make antihydrogen. Since electrons and positrons have opposite charges, their net charge is zero, and so a pair of them can annihilate, if they collide, into photons that carry off energy. Indeed, anytime a positron finds itself on the earth it soon hits an electron and annihilates. So there are normally few positrons around, except in outer space. Similarly, antiprotons annihilate with protons, so few antiprotons are around. We will return later to briefly consider why our world is mainly matter and not antimatter.

Neutrinos

In 1930, experiments seemed to suggest that some nuclear decays did not conserve energy—less energy was detected in the final state than in the initial one. Wolfgang Pauli proposed that the energy was being carried off by an unseen particle, one that had no electric charge and interacted only via the weak interaction, in which case it would pass right through normal detectors without any effect. Yet it would carry off energy. If that were true, certain predictions could be made for the motion of the other particles that emerged in the decay, and those were verified, so within a few years most experts were convinced from the indirect evidence that "neutrinos" indeed existed. It wasn't until 1958, almost thirty years after they were postulated, that they were directly detected. Neutrinos are often denoted by the Greek letter nu (ν).

More Quarks and Leptons

There are two more particles like the electron. All their properties are identical to those of the electron, except that they are heavier, and each is associated with its own neutrino. Although we don't yet know why that is so, the Standard Model theory still correctly describes the behavior of these particles. In experiments, various numbers of these particles are produced going in various directions. They are unstable particles, decaying in less than a millionth of a second into electrons, positrons, neutrinos, and photons, so they do not stay around after they are produced or end up in anything we see. In all cases the Standard Model can correctly predict the number of them, their directions, energies, lifetimes, and all other aspects of their behavior. One is called the muon (denoted by the Greek letter μ), and the other is called tau (τ). The electron, muon, tau, and their associated neutrinos, because they behave in very similar ways, are grouped together in a class of particles called leptons.

There are also two more quarks just like the up quark except that they are heavier. All the same remarks about predictions, lifetimes, and so on hold—the Standard Model correctly predicts all aspects of their behavior. They are called the charmed quark (denoted "c") and the top quark ("t"). And there are two quarks just like the down quark except that they are heavier. They are called the strange quark ("s") and the bottom quark ("b"). While the Standard Model can fully describe the behavior of all of these extra quarks and leptons, it cannot tell us why they are there. We'll see briefly later that using the framework of string theory does address why the quarks and leptons exist, and which ones.

Usually these quarks and leptons are grouped into three so-called families. The first family has the electron and the up and down quarks that together make up all we perceive, plus the neutrino associated with the electron. The second family has one of the particles like the electron but heavier (the muon), the neutrino associated with the muon, and quarks that are like the up and down quarks but heavier (the charm and strange quarks). The third family has an even heavier electron-like particle (the tau), the neutrino associated with the tau, and the other heavy quarks (top, like up, and bottom, like down). Good experimental evidence implies that there are no more such families of quarks and leptons. This apparent replication of particles into three groups, each group exactly identical to the others except for the increasingly heavy masses of the particles comprising each group, is one of the most surprising and disconcerting discoveries of the past century in particle physics—it is called the "family problem." We don't know yet why nature has three families instead of just one. As far as we know today, one family would have been sufficient to build our world. It isn't that we have not yet been clever enough to figure out from the Standard Model why there are three families but, rather, that the Standard Model does not contain principles that could explain this. We hope that a future extension of the Standard Model will; an attractive feature of string theories is that they appear to be able to address this issue, though we do not yet know if they do so correctly.

Higgs Boson(s)

The Standard Model predicted that another kind of particle should exist, a "Higgs boson." Experimental proof of its existence completes the Standard Model. The precise properties of the Higgs boson, and how many kinds of Higgs bosons exist, will give us clues to how the Standard Model will be extended; for example, supersymmetry predicts that at least five types of Higgs bosons exist. The existence and properties of Higgs bosons are a special test of the Standard Model because Higgs bosons are different from any previously known kind of particle. Leptons and quarks are basically all particles like electrons, just carrying different electric or weak or strong charges. Similarly, all the quanta that mediate the weak and strong interactions are like the photon. However, Higgs bosons are really a new kind of matter, predicted by the Standard Model to exist. Their successful prediction and discovery is a spectacular achievement of human reasoning. Elsewhere in the book I describe the implications of the discovery for improved understanding of the underlying theory that extends the Standard Model.

NEW IDEAS AND REMARKABLE PREDICTIONS OF THE STANDARD MODEL

The theory of the Standard Model is based on two major theoretical principles. The first is that the underlying laws of nature do not change when certain apparently different particles are interchanged in the equations. We say that the theory is invariant under those interchanges; such an invariance is an example of a symmetry. One such pair is the electron and its neutrino. These two have similar properties and behavior, but they differ in that the electron has an electric charge while the neutrino does not, and they have different mass. Imagine a kind of abstract space (not our real space) where particles can be represented by arrows. If a particular arrow points one way, it is an electron; if it points the opposite way, it is the electron neutrino. The charge is a label that tells which way the arrow is pointing at the moment. Particles don't change the direction they point when they are sitting around, but they can when they interact. Other possible interchanges include up quark ↔ down quark (i.e., interchange the symbols for the up quark and the down quark in the equations). Requiring that the equations do not change when the particles are interchanged puts powerful constraints on the form of the theory and on its predictions. One clear prediction is that if one particle that should belong to a hypothetical interchangeable pair is found, then the other must exist. The charmed quark and the top quark were both predicted to exist by this argument, and then were found with the expected properties. We don't yet know why this principle should hold for the theory that describes nature, but it does emerge as a prediction from some forms of string theory, which is encouraging. For those who want to keep track of the jargon, this invariance is a prediction of "non-Abelian gauge theories."

Another prediction based on this argument is that the particles must have the same mass if they can be interchanged, since the equations can depend on their mass. Some technical arguments show that this is possible only if all of them (quarks and leptons and W and Z bosons) have zero mass. Unfortunately, that does not agree with the data, since these particles do have mass, which is where Higgs boson physics enters the picture. It is the interaction with the Higgs boson that allows all particles to have not only mass but different masses. That interaction has a special form that allows the equations to retain their invariance under interchange of the particles, but still allows the particles to have (different) mass. The terms in the Lagrangian that represent the interaction with the Higgs boson are in the third and fourth lines on the T-shirt in Figure 2.1. Photons and gravitons and gluons don't directly interact with Higgs bosons, and therefore they remain massless. The existence of the Higgs physics is essential to make a consistent theory that can describe the actual particles, and to have a universe built from massive elementary particles.

The symmetry of the Standard Model under interchange of certain particles is well hidden. Nothing about the actual properties and behavior of particles explicitly shows the interchange behavior. It showed up only after clever physicists guessed it was there, and then it predicted some implications that could be tested experimentally. Once this was indeed demonstrated for the Standard Model, physicists began to take more seriously the possibility of other well-hidden symmetries in nature. In particular, that possibility made it easier to think about having a theory with a symmetry under interchange of bosons and fermions—that is, supersymmetry. (Bosons and fermions are defined explicitly later in this chapter.)

The second principle forming the foundation of the Standard Model is really a consequence of quantum theory rather than a new idea, but several decades passed before it was fully understood—and in any case, it emerged only as part of the Standard Model. It seemed like a new idea. If there is a particle, such as an electron, carrying a charge, then it is impossible to make a consistent quantum theory unless an additional field exists and interacts with the electron. That additional field has precisely the properties of the electromagnetic field, so it can be interpreted as being the electromagnetic field. Since the quanta of the electromagnetic field are photons, the photon must exist once electrons do, given that quantum theory provides the rules by which nature operates. Thus in the Standard Model the electromagnetic force and the photon are not an extra or separate part of the world, with electrons and photons happening to interact. Rather, once the electron exists, so must the photon. The existence of the photon is explained. With the Standard Model we finally understand what light is.

The same argument applies to the weak and strong charges. Once any particle carries a weak or a strong charge, there must exist particles like the photon but associated with the weak and strong forces. These particles are the W and Z bosons for the weak force, and the gluons for the strong force. The existence of these particles was actually predicted by this argument, and they were found in planned experiments with exactly the properties they were expected to have. Correctly predicting the existence of these particles was one of the strongest reasons why we are confident that the Standard Model is indeed here to stay.

Experimental Foundations of the Standard Model

A mere description of the Standard Model itself does not do justice to the complex and interesting experimental history and tests that both led to its formulation and gave us confidence that it indeed describes nature correctly. Without data of many kinds—the Lamb shift of hydrogen energy levels, inelastic scattering of electrons, parity violation, neutrino beams, pion decay, neutral currents, W and Z production and decay, charm production, and much more—the Standard Model would not have been developed. It is too far removed from our world to have emerged from theory alone. Actually the situation is more interesting. It is at least conceivable that much of the Standard Model could have been deduced. But almost no one would have believed the results.

There are lots of theories proposed. There would have been no consensus and no progress. The experiments are crucial for wide acceptance at each stage, and for continuous progress.

These and others were major crucial experiments, though I will not consider them further in this book. Doing those experiments required new developments in accelerator and detector physics, and computer technology, and the insights to carry out the right experiments, and the skills to get them right. Later I will discuss some of the experiments that might establish and test supersymmetry and even string theory, and get a better sense of the role of experimentation. Some of the books in the annotated bibliography describe the experimental foundations of the Standard Model.

Spin, Fermions, and Bosons

All particles have a property called spin. As with other features of the Standard Model, the term "spin" is best thought of as an analogy rather than as a property the same as that of a spinning top. We can fully describe the effects of having spin and every way that spin affects the behavior of particles. Spin is like the behavior a spinning particle would have, and it is another quantum kind of property. Particles can have only amounts of spin in certain chunks of a basic unit. The allowed values of spin are zero, half-integer, and integer multiples of the basic unit, ½, 1, and so on—the value of the unit doesn't matter for us.

A successful prediction of quantum field theory is that all electrons are identical to each other. All electrons anywhere are indistinguishable. And all photons are identical to each other, and all of each kind of particle is indistinguishable from others of that kind. This is not a puzzle, since all electrons are quanta of the electron field, and we would expect all quanta of a given kind of field to be identical. Being identical to the others doesn't matter much if we are only considering one of them, but when more are involved it has some surprising implications in quantum theory. The details don't matter for us; the important thing is that particles with integer spin (0, 1, etc.) behave differently from particles with half-integer spin (1/2, 3/2, etc.). The first set are called "bosons" and the second "fermions."

For an individual particle it doesn't matter much if it's a fermion or a boson. However, when lots of particles get involved, the behavior of fermions and of bosons is remarkably different. The existence of fermions is responsible for the stability of matter, whereas all of the force-mediating particles are bosons. Thus physicists normally think of fermions and bosons as very different kinds of particles. It was a great surprise that there could be a symmetry that could relate them, supersymmetry.

BEYOND THE STANDARD MODEL

In a sense the Standard Model has achieved the goals that science traditionally set. Finally, for the first time in history, we have a complete description of how our physical world works. There are no puzzles or significant conflicts between theory and experiment. It is the underlying physical theory for all of chemistry, the behavior of matter, the behavior of a star, and all that we see. As explained earlier, that does not at all imply that doing science has ended, because the equations still need to be solved for many systems, and unexpected properties emerge for complex systems (see Chapter 3). Once we had the Standard Model we could ask new questions about the universe as a whole, and we could start to ask questions about why the Standard Model took one form and not another. Many questions that were philosophical or entirely speculative have recently become normal research topics in physics, astrophysics, and cosmology. In this section I will list some of those topics. They are listed partly to provide an overview of why so many theorists are confident that new physics that extends the Standard Model (such as supersymmetry) is just around the corner.

- We saw earlier that in order to describe the masses of the particles, the Standard Model had to contain some new interaction that modified the particle-interchange symmetry of the Standard

Model in a very specific way—and the Higgs physics could indeed do that. The Standard Model itself *cannot* provide a mechanism that produces or explains the Higgs physics in the needed way, so some new physics beyond the Standard Model is essential. Supersymmetry can provide that new physics. At present, it is the best known way to do it. It can *explain* how the interactions of particles with the Higgs field can provide mass and allow different mass for different particles.

- As also mentioned above, if we effectively look at the weak, electromagnetic, and strong forces at smaller distances by using quantum theory to magnify them, we find that they look increasingly similar to each other, though there is no apparent reason why that should be expected to happen. In the supersymmetric extension of the Standard Model, the forces essentially become identical at sufficiently small distances, strongly encouraging optimism about the possibility of truly unifying them.

- Planets further out in the solar system take longer to orbit the sun, in accord with Kepler's laws, discovered four centuries ago. When the motion of stars in our galaxy or of galaxies in clusters of galaxies is examined, it is found that Kepler's laws do not seem to apply if they are written to include only the visible stars, but they do seem to apply if a large amount of nonvisible matter (i.e., "dark" matter) is included. If we couldn't see the moon, we could deduce from the tides that it was there once we understood how gravity worked. Similarly, we can deduce that the dark matter is there. When this was done on all scales in the universe, we learned that far more dark matter is present than could be accounted for by Standard Model particles. Many possibilities for the dark matter have been suggested; the lightest of the new superpartners predicted by supersymmetry is the favored one. Since we don't yet know which superpartner it is, it is often called the LSP, for lightest superpartner. We will return to this question in Chapter 6 and consider how we can learn experimentally what actually makes the dark matter. "LSP" is one of the few acronyms used in this book.

- People and planets and stars are made of protons and neutrons, but not of the antiparticles of protons and neutrons. The Standard Model cannot provide a mechanism to explain how to start with a universe that initially is symmetric between particles and antiparticles (as we would expect from the big bang) but evolves to give the observed asymmetry. Notice that this is a case where we do not know if we can find an explanation: one could simply say that however the universe began, it started with an excess of matter over antimatter—but to most physicists that is philosophically unacceptable. One can imagine understanding a universe that began with equal numbers of particles and antiparticles, with no net number of either, but not a universe that somehow had more particles than antiparticles from the beginning. In fact, the famous physicist and Soviet dissident Andrei Sakharov was the first to point out, in 1967, that if certain conditions held it would be possible to explain how such an asymmetry could evolve from initial symmetry as the universe aged. Those conditions can indeed hold if the Standard Model is extended in certain ways. There are actually several alternatives, and most of them rely on supersymmetry. We return to this issue in Chapter 9.

- The Standard Model is a consistent theory whether or not neutrinos have mass. Technically neutrinos with mass could be described in the Standard Model the same way that electrons are, using the Higgs interaction. But experiments show that neutrinos have at most about a million times less mass than electrons, and it would not make sense for their mass to arise the same way as the electron mass yet be much smaller. There is recent experimental evidence that at least two neutrinos have nonzero but very small mass. Since the late 1970s theorists have been inventing ideas that would lead to theories in which neutrinos have mass. Since the middle 1980s all experts in this area have agreed that it is extremely unlikely that neutrinos are massless, and that all approaches to incorporate a mass for them in the theory require some new physics beyond the Standard Model. Here supersymmetry is helpful in formulating ideas, but it has not played an essential role so far. Surprisingly, all approaches to include neutrinos in the mathematical theory of the Standard Model require introducing a new mass scale into the theory besides the masses of the Standard Model itself or the Planck mass. If the underlying theory is supersymmetric, the needed extension of the Standard Model to include neutrino masses can be reliably done.

- The phenomena listed so far—the need for a mechanism to account for Higgs physics, the matter-antimatter asymmetry, neutrino masses, and dark matter—are based on real data that we can hope to understand better with an extended theory that includes the Standard Model. We can also list a number of "why" questions that we can hope to understand. Why are the basic particles quarks and leptons rather than something else? What *are* quarks and leptons? Why is the Standard Model theory unchanged if we interchange electrons and their neutrinos, or up and down quarks, but not if we interchange (say) electrons and muons, or electrons and up quarks? To put it more generally, we can ask why the Standard Model is what it is. There were other reasonable and elegant theories that addressed the same issues as the Standard Model, but data showed that the Standard Model was the correct one, the one that actually described nature. Why are there three families of quarks and leptons? Why is it so hard to make a quantum theory of gravity? Do the different forces in fact become indistinguishable and somehow turn into one force at very small distances, so that we can have a simple picture of the universe as governed by one underlying force that just appears as several different ones when we observe it at our large distances? More of these kinds of questions could be listed—but this list suffices to make the point. We have no guarantee that we can ever answer such questions. But we can try, and in fact research so far has in every case suggested possible answers, sometimes in the context of string theory (see Chapter 8). In every case, supersymmetry is relevant to the answers. I will discuss some of these possible answers in the following chapters.

Readers unfamiliar with particle physics may have been amused by, or had some other reaction to, the largely metaphorical terminology: up and charmed and top quarks, gluons that bind, families, strings, and so on. Each of these terms has a precise meaning and corresponds to specific quantities in equations. They do not correspond to things in our everyday world, but we need to talk about them. Particle physicists have chosen to humanize them with familiar names that usually carry a reminder of the role of the particle or interaction we are referring to. The names often help us to think metaphorically about the way the world works at tiny distances. Sometimes people misinterpret the apparently cavalier terminology as an irreverent one, but in fact it shows great respect for these new properties of the world that have been discovered as we probe nature in new ways. Readers who would like more information about the Standard Model can turn to the list of recommended books and articles at the end of this volume.

An interesting perspective on physics beyond the domain of the Standard Model is provided by asking about the relative status of experiment and theory as a field progresses. Particle physics began as a field about a century ago: one could date it to the discovery of the electron, or perhaps a few years later to the use of particle beams, which allowed us to learn that the atom had a nucleus. From then until the discovery of the Standard Model in the early 1970s experiment was ahead of theory, with many unexpected discoveries and essentially no successful predictions. Once the Standard Model was written, it explained many puzzles and made many dramatic and successful predictions. There were no experimental surprises since then—all results were either clearly predicted, or at least anticipated. Once we go beyond the domain of the Standard Model, the situation is mixed. Today theory and experiment are again making progress together, as they did throughout most of the history of physics. But the situation is different in an important way from that of the century before the Standard Model, because now the Standard Model provides a framework allowing new kinds of questions to be formulated. The Higgs boson was predicted by the Standard Model except for its mass. Physics beyond the Standard Model strives to do more in several areas.

3
· · · · · · · · ·

Why Physics Is the Easiest Science:
Effective Theories

IF WE HAD TO UNDERSTAND the whole physical universe at once in order to understand any part of it, we would have made little progress. Suppose that the properties of atoms depended on their history, or whether they were in stars or people or labs—we might not understand them yet. In many areas of biology and ecology and other fields, systems are influenced by many factors, and of course the behavior of people is dominated by our interactions with others. In these areas progress comes more slowly. The physical world, on the other hand, can be studied in segments that hardly affect one another, as I will emphasize in this chapter. If our goal is learning how things work, a segmented approach is very fruitful. That is one of several reasons why physics began earlier in history than other sciences and has made considerable progress: it really is the easiest science. (Other reasons in addition to the focus of this chapter include the relative ease with which experiments can change one quantity at a time, holding others fixed; the relative ease with which experiments can be repeated and improved when the implications are unclear; and the high likelihood that results are described by simple mathematics, allowing testable predictions to be deduced.)

Once we understand each segment we can connect several of them and unify our understanding, a unification based on real understanding of how the parts of the world behave rather than on philosophical speculation. The history of physics could be written as a process of tackling the separate areas once the technology and available understanding allow them to be studied, followed by the continual unification of segments into a larger whole. Today one can argue that in physics we are finally working at the boundaries of this process, where research focuses on unifying all of the interactions and particles; this is an exciting time intellectually. In recent years physicists have understood this approach better, and made it more explicit and formal. The jargon for the modern way of thinking about theories and their relations is the method of "effective theories." I will use this bit of terminology in the present chapter and occasionally later in the book. Sometimes this approach has been described as "reductionist"—a word that may have different meanings and implications for different people, and negative connotations, so I won't use it here. For physicists, however, "reductionist" implies simultaneously separating areas to study them *and* integrating them as they become understood. This chapter aims to help understand where the Higgs boson discovery and supersymmetry fit into a broader picture and, more generally, to understand better how scientific research works.

Organizing Effective Theories by Distance Scales

Probably the best way to organize effective theories is in terms of the typical size of structures studied by a particular effective theory, which we speak of as the distance scale of normal phenomena described by that theory. Imagine starting by thinking about the universe at very large distances, so large that our sun and all stars look like small objects from such distances. This is the effective theory of cosmology, where stars cluster in galaxies because of their gravitational attraction, and galaxies are attracted to each other and form clusters of galaxies. Because of gravitational attraction everything is moving, on a background of the expanding universe. The only force that matters is gravity. We can use simple Newtonian rules to describe motion—deviations due to effects described by quantum theory are tiny and can be ignored. We can study how stars and planets and galaxies form, their typical sizes, how they distribute themselves around the universe, and so on. It doesn't matter whether the particles that make up stars and planets are made of quarks or not, or how many forces there are at the small distances inside a nucleus. The large-scale universe is insensitive to what its contents are except for their mass and energy. Because of this indifference, cosmology can make progress regardless of whether we understand how stars work, or whether protons are made of quarks, and so on. We can learn from astronomy data that there is dark matter. However, this also implies that if the dark matter is composed of particles we cannot learn from astronomy or cosmology what kind of particles they are, because cosmology is largely insensitive to the properties that distinguish one particle from another, such as their masses and what charges they carry.

Next consider smaller distances, about the size of stars. We can study how stars form, how they get their energy supply, how long they will shine—we can work out an effective theory of stars. While doing that we can ignore whether they are in galaxies, or whether there are top quarks or people. Let's go to a smaller distance—say, the size of people—and consider the physics. Gravity keeps us on the planet, but otherwise it is the electromagnetic force that matters. All of our senses come from mechanical and chemical effects based on the electromagnetic force. Sight consists of photons interacting with electrons in our eyes, followed by electrical signals traveling to our brains. Touch begins with pressure affecting cells in the skin, leading to electrical signals propagating to the brain. Hearing starts with air molecules hitting molecules in the eardrum, interacting via electromagnetic forces. Friction, essential for us to stay in place or to move, is due to electromagnetic forces between atoms. We don't need to know about the weak or strong forces, or the galaxy, to study people-sized physics. The sun inputs energy to the earth, providing all of our food and essentially all of our energy, mostly from stored solar energy; but otherwise how the sun works does not matter. Here is a case where phenomena from one effective theory provide input to another in a very specific way—the earth can be viewed as a closed system except for the input of solar energy. From the point of view of the effective theory of the earth, how the sun generates its energy is irrelevant.

Now consider atomic size. Here we'll be able to see even better how powerful the effective-theory idea is. To be able to use the basic equations that govern atoms we have to input some information, essentially a few properties of the electron (its mass and electric charge and spin), and the same properties for each of the naturally occurring nuclei if we want to describe the whole Periodic Table of the Elements. A description of atoms doesn't need to take account of whether stars or galaxies or people exist, nor if the nucleus is made of protons and neutrons, nor if protons are made of quarks.

Before we go to even smaller distances, this is a good place to stop and consider some of the implications of this way of thinking. When we have a tentative theory of atoms, we want to test its predictions and whether it explains phenomena we already know. The predictions for the lifetimes of excited states of atoms, for the energies of photons emitted by atoms, for the sizes of atoms, and so on depend on pieces of input information—on the masses and charges and spins of the electron and the nuclei. Every effective theory has some input parameters such as these. Without input information about the electron and nuclei, we could solve the equations but not evaluate the results numerically, so we could not test the theory or make any useful predictions, or know if we had the right theory. For example, the radius of the hydrogen atom is $h^2/e^2 m_e$ (h is Planck's constant, m_e is the electron's mass, and e is the magnitude of the electric charge of the electron). If we measure the size of the hydrogen atom, we still can't check whether the theory is working unless we know (by measurement or calculation) the mass and charge of the electron and Planck's constant (all three were measured nearly a century ago). Note that the size of the atom gets larger if the electron mass is smaller (since it is in the denominator in the expression for the radius above), and the radius would become infinite if the electron mass were zero, as it would be without its interaction with the Higgs field.

Now an important aspect of effective theories can be emphasized. When we study the effective theory of the nucleus, we want to be able to calculate the masses and spins and charges of the nuclei—those parameters are results derived in the effective theory at that level. But for the next higher level, the atom, those parameters are the input. They can be input if they are measured, whether or not they are understood in the effective theory of nuclei. Similarly, the electron is a fundamental particle whose mass and electric charge will, it is hoped, be calculable someday in a fundamental theory such as M-theory or string theory, but for the effective theory of atomic physics it does not matter whether they are calculable or understood if they have been measured. We have known the numerical value of the mass of the electron since the beginning of the twentieth century, but we do not yet understand why it has that value. The value of the electron's mass can be input into every effective theory that depends on the electron. Every effective theory so far has some input that, for that effective theory, is a "given"—not something to be questioned, such as the mass of the electron for the effective theory of atoms.

If a theory has inputs, it is an effective theory. From this point of view the goal of particle physics is to learn the ultimate theory at the smallest distances, recognizing it as the theory for which no parameters have to be input to calculate its predictions. For the ultimate theory it is not satisfactory to input the electron mass; rather, it is necessary to be able to calculate it from basic principles and to explain why it has the value it does.

Every effective theory is based on others: it is effective theories all the way down, until the ultimate one, which has been called the final theory. Each effective theory has certain structures, which bind together at its level—stars, atoms, nuclei, protons. Stars are made of nuclei and electrons bound by the gravitational force as viewed from the effective theory of stars, but for the effective theory of cosmology they are just inputs characterized by a mass and brightness. Nuclei are bound states of neutrons and protons for the effective theory of nuclei, but for the effective theory of atoms they are merely point-like inputs. All systems and structure are inputs at one level of effective theories, but something to be derived and explained by the effective theory at a smaller distance. In a sense, a given effective theory can be explained in terms of shorter distance theories and the input from them. Dirac said that his equation that unified special relativity and quantum theory for the interactions of electrons and nuclei explained all of chemistry, and in a sense he was right. His equation, plus the input parameters describing electrons and nuclei, in principle explained all chemical processes. In another sense he was not right, because in practice one could never start from the Dirac equation and calculate the properties of molecules, or figure out their structure or how to construct new molecules with certain desired properties—the questions are just too complicated to solve. For example, Dirac could not have deduced that water is wet from his equation. For each effective theory new regularities or laws are found, and properties arise that are not predictable *in practice.* They are often called emergent properties. Life is an emergent property. Physics tells us everything that molecules can do and cannot do. In particular, physics can tell us that life will not emerge on some planets if circumstances are too adverse, but it cannot guarantee that life will emerge on a planet where conditions are favorable, even though that may be very probable.

Another way to view effective theories is in terms of types of understanding. At its own level, an effective theory provides a "how understanding," a description of how things work. But for the effective theory above it, at larger distances, the smaller-distance effective theory explains all or some of the input parameters, thus providing a "why understanding." For example, nuclear physics describes the properties of nuclei, using the proton and its electric charge, spin, mass, and magnetic properties as given, unexplained input. But the Standard Model provides the explanation, allowing the calculation of all those properties of the proton in terms of quarks bound by gluons.

Yet another perspective appears if we observe that in general all areas of science are intrinsically open-ended—chemistry, the physics of materials, geology, biology, and so on. There is no end to the number of possible systems and variations that can be studied. But particle physics and cosmology are different. If the fundamental laws that govern the universe are found and understood, and the inputs are calculated in that theory, that's it—these two fields (which are merging into one) will end.

Having examined some implications of effective theories, let's return to the progression to smaller distances. We can go from atoms through the effective theory of nuclei, protons, and neutrons, to quarks and leptons. Another important point is that each effective theory works well at its level, but it breaks down as we go to smaller distances and find new kinds of structure. When we went inside protons we found quarks, so we could not make a theory of protons unless we understood quarks and their interactions as well.

To go deeper into matter, we will find it helpful to keep track of the distance scales numerically. Since we will cover a huge range of distances, we need to use powers of ten (remember that each step in the power is a factor of ten in the result): 10^{-1}\$ is a dime, and 10^{-2}\$ is a penny. 10^3\$ is one thousand dollars. There are two scales that are useful for us to keep track of, meters that are a typical human size, and another length called the Planck length, after Max Planck who first introduced it soon after he took the initial step toward quantum theory in 1899. The Planck length is extremely small—we'll understand it better later in this chapter. People are typically a meter or two in size. All of our analysis will be very approximate, so we won't worry about whether we talk of the height or width or radius of a system. We'll keep track of powers of ten but not worry about distinguishing between things that might differ by a factor of two or so in size. People are about 10^{35} Planck lengths (1 followed by 35 zeros) tall, 1 or 2 meters. Atoms are about 10^{-10} meters (one ten-billionth of a meter)—about 10^{25} Planck lengths—in radius. Protons are about 100,000 times smaller than atoms, 10^{-15} meters or 10^{20} Planck lengths. Many particle physicists currently expect that quarks, leptons, photons, W and Z bosons, and gluons will ultimately be understood as having a string-like extension if we could view them at a distance scale of about 1 Planck length or 10^{-35} meters (a decimal point followed by 34 zeros and a 1); they should seem point-like until we can study them at that scale.

In the language of this chapter we can think of the Standard Model (see the previous chapter) as the effective theory of quarks and leptons interacting on a scale of about 10^{-17} meters, or 10^{18} Planck lengths, about 100 times smaller than protons and neutrons. Sometimes we call this the "collider scale" since it is associated with the typical energies at which the experimental collider facilities operate.

The goal of particle physics is an ultimate theory of the natural world. What should we call it? People have called it a Theory of Everything, but that name is somewhat misleading since it is really not a theory of weather, stars, psychology, and everything at once. Steven Weinberg called it the final theory. That's a good name, but it can be misinterpreted as the last in a succession of theories that replaced each other, as if all the theories on the way to the final one should be discarded. In fact, all the effective theories coexist simultaneously; all are part of our description of nature. A good name would be the "primary theory," following Lucretius, suggesting the theory one arrives at after going through a sequence of effective theories at smaller and smaller distances. However, on balance, sticking with "final" might be best for communicating with people easily, so I will do that. As we will see more clearly in a few paragraphs, the final theory should be the description of nature at a distance scale of about 1 Planck length, or about 10^{-35} meters. How can we journey the many orders of magnitude from the Standard Model to the Planck length?

SUPERSYMMETRY IS AN EFFECTIVE THEORY, TOO

If nature is indeed supersymmetric, one of the wonderful bonuses we may get is a way to carry out the journey through those orders of magnitude. Supersymmetry is an effective theory too, but it may be the penultimate one that will take us from the Standard Model to the theory near or at the Planck scale. Supersymmetry is an effective theory because it still needs some input parameters to describe the masses and interactions of the particles; those inputs should be predictable by the theory near the Planck scale (e.g., M/string theory). Qualitatively, the supersymmetric Standard Model should become the effective theory at distances of 10^{-17} to 10^{-18} meters and remain the effective theory down to nearly the Planck scale. It has special properties that allow it to cover that large range, rather than breaking down at shorter distances as most effective theories do.

In the past we have been able to do the experiments that were essential to make progress as the technology developed and allowed us to probe more deeply. The Planck length is too small—there will never be direct experiments possible at that scale. This statement is not just an extrapolation based on current technologies or costs. It is not just a matter of getting higher energy probes: the probes have to have the energy concentrated into a region smaller than the scale of interest, and before we can do that we not only run into limits like cost, we run into natural limits. Nevertheless, there are a variety of ways to test ideas about Planck-scale physics (this is explained further in Chapter 9). Saying we cannot test Planck-scale physics is like saying we cannot test the big bang theory. Even though there was no one present at the big bang, there are relics that provide convincing tests that it occurred. We already have some indirect ways to test ideas about physics at the Planck length, but supersymmetry will allow us to add many systematic tests. It will give us techniques to take a prediction at the Planck length and calculate what is predicted at the distances colliders can probe in the coming years (about 10^{-18} meters) or to take data from colliders and calculate the form of the theory at the Planck length implied by the data. With supersymmetry we will be able to test ideas about M/string theories (see Chapter 9) or whatever form the final theory may take—without it we do not know how to do that. Of course, that bonus does not guarantee that nature is indeed supersymmetric, but it is a powerful motivation to study the theory and do the experiments needed to find out.

THE PHYSICS OF THE PLANCK SCALE

Whenever we describe a segment of nature we have to talk about the actual quantities that are calculated or predicted or explained in units—meters, or seconds, or kilograms or other appropriate units. In Carl Sagan's novel *Contact* a signal from an extraterrestrial intelligence has been detected, with instructions on how to build a machine to facilitate communication. There is a conversation between a scientist and an administrator:

> "Don't ask why we need two tons of erbium. Nobody has the faintest idea."
> "I wasn't going to ask that. I want to know how they told you how much a ton is."
> "They counted it out for us in Planck masses. A Planck mass is—"
> "Never mind, never mind. It's something that physicists all over the universe know about, right? And we've never heard of it."

For every effective theory there is a natural system of units, one in which the description of phenomena is simple and not clumsy. It would be silly to measure room sizes in Planck lengths just because the final theory is best talked about in those units. Consider the units for atoms more closely. The radius of an atom can be expressed in terms of the properties of the electron plus Planck's constant h, which sets the scale of all quanta. (Planck's constant is the fundamental, universal constant of quantum theory.) If we denote the electric charge of the electron by e, and the mass of the electron by m_e, we find that the radius (R) of the hydrogen atom, the simplest atom, is $R = h^2/e^2 m_e$. The size of the atom is fully determined by these inputs. Nothing else matters—the nucleus, for example, is just a tiny object at the center. Once we know R, we can express the sizes of all atoms in terms of R; we don't need to use the input of h or the electron properties any more. R is the natural size unit for atoms. Atoms with different numbers of electrons will have somewhat different sizes, with radii such as $1.2R$ or $2.4R$, but all will be some number that is not too big or small times R. R is expressed in terms of parameters that are givens for atomic physics; we hope e and m_e can be calculated someday in M/string theory, but they cannot be understood by atomic physics.

We can learn a great deal from this kind of analysis. For example, this expression for the size of an atom has major implications. It tells us that the size of atoms is essentially a universal quantity. Given the basic quantities (Planck's constant and the mass and charge of the electron), the size of all atoms of all kinds, anywhere in the universe, is determined. Since mountains and plants and animals are all made of atoms, their sizes are approximately determined by the size of atoms and the electromagnetic and gravitational forces. Combining atoms into genes and cells to evolve an organism that can manipulate and deal with the world requires a large number of cells and sets a minimum size for the organism. Having a brain with enough neurons to make enough connections to make decisions about the world requires a minimum-sized brain, since the atoms cannot be made smaller than the size determined by the radius R. Nothing the size of a butterfly will be able to think, anywhere in the universe. Thinking organisms could be much larger than people, but they do not need to be, so it follows from general principles that all intelligent life is expected to be about our size, not much larger or much smaller.

Suppose now that we have just discovered the final theory. To present the results we have to express the predictions and explanations in appropriate units. What units should we use? We expect the natural units for the final theory to be very universal ones, not dependent on whether the universe has people or stars. There is only one known way to make universal units. There are only three universal constants in nature common to all aspects of nature, all interactions, and all particles. They are Planck's constant h; the speed of light (denoted by c), which is constant under all conditions; and Newton's constant G that measures the strength of the gravitational force. Since Einstein proved that energy and mass are convertible into one another, and gravitation is a force proportional to the amount of energy a system has, everything in the universe feels the gravitational force. In fact, using these three quantities—h, c, G—we can construct combinations that have the units of length, time, and energy. We expect all the quantities that enter into the final theory or are solutions of the equations of the final theory to be expressible in terms of the units constructed from h, c, G. [For the interested reader, the result for the Planck length is $(Gh/c^3)^{1/2}$, which, as noted earlier, is about equal to 10^{-35} meters. For completeness, the Planck time is $(hG/c^5)^{1/2}$, which is about 10^{-44} seconds, and the Planck mass is $(hc/G)^{1/2}$, which is about 10^{-8} kilograms.] The Planck distance and time are extremely small, while the Planck mass (or equivalently energy) is very large for a particle.

Max Planck understood fully a century ago the universality of those units we call the Planck units. He wrote in his book *The Theory of Heat Radiation* (reprinted by Dover Publications, 1991):

All the systems of units which have hitherto been employed . . . owe their origin to the coincidence of accidental circumstances, inasmuch as the choice of the units lying at the base of every system has been made, not according to general points of view which would necessarily retain their importance for all places and all times, but essentially with reference to the special needs of our terrestrial civilization. Thus the units of length and time were derived from the present dimensions and motion of our planet. . . . In contrast with this it might be of interest to note that . . . we have the means of establishing units of length, mass, time . . . which are independent of special bodies or substances, which necessarily retain their significance for all times and for all environments, terrestrial and human or otherwise, and which may, therefore, be described as "natural units." The means of determining the units of length, mass, and time . . . are given by the constant h, together with the magnitude of the velocity of propagation of light in a vacuum, c, and that of the constant of gravitation, G. These quantities retain their natural significance as long as the law of gravitation and that of the propagation of light in a vacuum [and quantum theory] remain valid. They therefore must be found always the same, when measured by the most widely differing intelligences according to the most widely differing methods.

(I have left out some words to make this read smoothly, replacing them with ellipses, and since Planck didn't then know about the completion of the development of quantum theory I have added that term in brackets as he would presumably have included it.)

The Planck length and time can also be interpreted as the smallest length and time that we can make sense of in a world described by quantum theory and having a universal gravitational force. The arguments that teach us that are interesting and not too complicated, but to explain them we have to recall the definition of a black hole. Basically the idea of a black hole is simple. Imagine being on a planet and launching a rocket. If you give the rocket enough speed, it can escape the gravitational attraction of the planet and travel into outer space. If you increase the mass of the planet, you have to increase the speed needed to escape. If you increase the mass so much that the required speed exceeds the speed of light, then the rocket can't escape since nothing can go faster than light. The rocket and everything is trapped. Light also feels gravitational forces, so beams of light are trapped too. Since gravitational forces increase with decreasing distance, if you pack some mass into a sphere of smaller radius it is harder to escape from it, so the condition for having a black hole depends on both the amount of mass and the size of the sphere you pack the mass into.

A fascinating thing is that if we put an object having the Planck energy in a region with a radius of the Planck length, we satisfy the conditions to have a black hole! We cannot separate such a region into parts, or get information out from a measurement, so we cannot define space to a greater precision than the Planck length! Since distance is speed × time, and speed can be at most the speed of light, and there is a minimum distance we can define, there is also a minimum time we can define—that comes out to be the Planck time. We saw above that the

Planck scale provides the natural units for expressing the final theory when the units are constructed from the fundamental constants h, c, and G. Now we see a second reason for expecting the Planck scale to be the distance scale for the final theory: there does not appear to be a way even in principle to make sense of smaller distances or times. The times when events occur cannot be specified, or even be put in order, more precisely than the Planck time.

There is a third interesting argument that gives the same answer. The gravitational force between two objects is proportional to their energies, and grows as the distance between them decreases. Consider, for example, two protons. Normally the repulsive electrical force between them is much larger than the attractive gravitational force. But if the energies of the protons are increased to the Planck energy, then the gravitational force between them becomes about equal to the electrical force between them. All the forces become about the same strength at the Planck scale, rather than being widely different in strength as they are in our everyday world. Thus we might expect the gravitational force to unify with the others at the Planck scale, just as one might hope for in the final theory.

The arguments of this chapter have led in several ways to the idea that it makes sense to analyze the physical world with effective theories organized by the distance scale to which they apply, and to move toward a final theory that unifies the forces and particles and is valid at the smallest scale that makes sense, the Planck scale. Of course, these arguments do not prove that this is how nature works; we will not know that until we achieve such a description. The Planck scale is very small, but not beyond our imagination. Some readers may recall the delightful *Powers of Ten* book and movie produced by the designers Charles and Ray Eames, in which the universe was looked at in snapshots each ten times smaller than the previous one, starting with the largest cosmological distances. When the Eameses did this work, shortly before the Standard Model was discovered, they could not meaningfully go to smaller distances than the proton. Today the Standard Model takes us nearly three powers of ten smaller than the proton. From the universe down to the Standard Model domain is about 46 powers of ten, and from the Standard Model to the Planck scale only about 16 more powers of ten. Looked at that way, perhaps it does not seem so far.

THE HUMAN SCALE

Since the time of Copernicus, who taught us that the earth was not at the center of the universe, we have learned that if we want to understand the world we have to go beyond how the world seems to be and ask for evidence of how it is. It looks like the sun rises, but actually the earth orbits the sun. We have learned that matter in the heavens and matter on earth obey the same natural laws, that we are made of the same atoms as the earth and the stars, that we and all organisms on the earth evolved from cells, that we have unconscious minds that affect our behavior, that our star is only one of a hundred billion stars in our galaxy. We have learned that the rules that govern nature (quantum theory and special relativity) are not apparent in our everyday classical world, and that the laws of nature have symmetries that are hidden from us but are important (such as the particle-interchange symmetry of the Standard Model). Perhaps even the number of space dimensions of the world will be larger than the three of which we are aware. That may seem surprising, but so is the fact that the earth orbits the sun. To understand the universe, we must recognize that additional hidden aspects of nature may arise at scales far different from the human scale and learn how to uncover them. Supersymmetry is such a hidden aspect of nature.

4

· · · · · · · · ·

Supersymmetry and Sparticles:
What Supersymmetry Adds

SUPERSYMMETRY IS THE SURPRISING IDEA, or hypothesis, that at the deepest level, for the ultimate or final theory, the laws of nature don't change if fermions are transformed into bosons and vice versa. If nature is indeed supersymmetric, the *equations* of the final theory will remain unchanged even with that interchange of the particles. This should be so despite the apparent differences between how bosons and fermions are treated in the Standard Model and in quantum theory. The details of those differences don't matter for us.

Whenever an extension of the existing description of nature is proposed, it must pass many tests before it has a chance of being valid. It must be consistent with the rules of quantum theory and special relativity, and it must not change any of the tested consequences of the Standard Model. The fermionic or bosonic nature of particles comes from their spin, and spin is related to quantum theory and special relativity, both of which in turn involve space and time in their formulation. Thus the formulation of supersymmetry must also involve space and time as well as interchange of bosons and fermions. It is remarkable that such a property can be introduced without coming into conflict with some already established experimental or theoretical result.

The reader may be underwhelmed. Who cares if the final theory is invariant when its fermions and bosons are interchanged? OK, it's remarkable that this can be done in a relativistically invariant quantum theory, but so what? Actually, that's the way physicists originally felt about it too, for the most part. The supersymmetric theory was technically very beautiful, so theorists enthusiastically studied it, but it was spoken of as a "solution in search of a problem." I got into the field because I asked at a number of talks in the late 1970s how we would know if nature were actually supersymmetric or not, and the typical answer was that no one had thought very much about that. It was such a beautiful theory that we began to feel strongly that it was important to figure out how to test it experimentally.

There is indeed reason to be excited, but it wasn't obvious at the beginning. That's related to how supersymmetry arose as an idea, in a manner different from other ideas in the history of science. New ideas, including the Standard Model, had always come as a response to trying to understand observed regularities, or to puzzles, or to a desire to explain the behavior of aspects of the natural world, or to apparent inconsistencies in existing descriptions. Supersymmetry on the contrary was originally noticed as a property of certain models (actually, with fewer than three space dimensions) being studied for their own sake. A fascinating aspect of the history is that supersymmetry was not introduced to solve any experimental mystery or theoretical inconsistency. Over a decade as it was studied and better understood, theorists realized that it actually could solve a number of the important mysteries in particle physics and provide new approaches to others. For many physicists, that supersymmetry solved problems it was not introduced to solve was itself a powerful hint that it was indeed part of the description of nature. No other major physics concept has ever been introduced as an idea not related to any data or inconsistency and then found later to solve important problems.

In subsequent chapters we will examine some of these mysteries and their solutions in a supersymmetric world. For now, however, they are listed with minimal explanation to give an overview of the impressive

Figure 4.1
Copyright © 1994 by Sidney Harris

impact of the supersymmetric theory. Thus far the word "mystery" has been used in the detective story sense. We expect mysteries to have solutions, to be solved—usually by the end of the book. The following list is meant to indicate why many physicists work on the theory of supersymmetry or on testing it experimentally, and why nearly twenty thousand papers on supersymmetry have been written before there is explicit evidence that it is really how nature works.

- In order to incorporate the masses of the particles into the Standard Model description, physicists postulated the existence of a Higgs field. They also assumed it interacted in a very specific and somewhat enigmatic way. The Higgs physics was a mysterious part of the Standard Model, hard to accept and hard to test, though technically it solved the problem for which it was introduced. In particular, some new physics beyond the Standard Model *must* exist to provide the Higgs physics; the Standard Model simply cannot provide an explanation of the Higgs interaction needed to account for the masses of particles in a consistent way. Then, in 1982, several theorists figured out that if the Standard Model was extended to be supersymmetric it could provide an elegant physical explanation for the Higgs physics. For many theoreticians this was the result that convinced us that supersymmetry was a property of nature, not just nice mathematics. Even better, in order for the supersymmetric approach to Higgs physics to work, it was necessary that the top quark (whose mass was measured as recently as the 1990s) be unusually heavy compared to the other quarks and leptons. That the top was predicted to be heavy and was indeed confirmed to be heavy a decade later by data was a powerful indirect test of the validity of supersymmetry.

- The Standard Model by itself has a very serious conceptual problem. We saw in Chapter 3 that the natural scale for the final theory was the Planck scale, about 10^{-35} meters. The Standard Model is a description of quarks and leptons and their interactions, at a scale of about 10^{-17} meters. The problem is that, in a quantum theory, physics at every scale may contribute to physics at every other scale, so it may not be consistent to have these two scales so separate: the Standard Model scale and the Planck scale should be very near each other, and in fact the Standard Model scale should be near the Planck scale. Another way to view this problem comes from recognizing that in the Standard Model all the masses of electrons, quarks, and Ws and Zs should be either zero or about the Planck mass. For the Standard Model this is indeed a major problem, even though it is a conceptual problem that does not explicitly affect its experimental predictions. The problem has two parts. First, given that there is a separation of the Standard Model scale from the Planck scale, why does the Standard Model end up where it is (at about 10^{-17} meters) and not at some other scale? Second, and more important conceptually, what can make the theory maintain that separation in a mathematically consistent way? The supersymmetric Standard Model solves the second problem and gives insight into the first. It does so in a manner that uses the unification of fermions and bosons in an essential way. The very nature of fermions and bosons implies that they contribute to the coming together of scales in ways that cancel, so the mixing of scales can be cancelled in a general way and thus can solve this problem. Technically, the rules of quantum field theory make the contributions have opposite signs and cancel to a good accuracy.

- For two centuries, physicists have been actively trying to unify our description of the forces of nature. Having five different forces, rather than one basic force, suggested we were missing some unifying principles. Maxwell succeeded in unifying electricity and magnetism, and the Standard Model unifies the description of weak interactions with electromagnetism, so there has been some progress. In a quantum theory one can calculate how a force would behave if one could study it as smaller distances. Remarkably, when this is done in the Standard Model for the electromagnetic, weak, and strong forces, it is found that they become more and more like each other at shorter distances, though finally they do not quite become equal in strength at any distance. More remarkably, when the study is repeated with the supersymmetric Standard Model, as was done in the early 1980s, it is found that the forces do become essentially equal at a very small distance, about 100 times larger than the Planck scale. That did not have to happen—nothing in the Standard Model implies that the forces should become equal. And if they do become equal, nothing in the Standard Model implies that this should happen at just the distance scale that would be expected if full unification occurred at the Planck scale. Presumably these are important clues toward a better understanding. Since we do not know the form of the complete theory at such small distances very well, we do not know how to extrapolate to even smaller distances, but some reasonable ways of doing that suggest that these forces become equal to the gravitational force at about the Planck scale, encouraging very much the idea that indeed the goal of understanding the forces of nature in a simple way will be reached. The superpartners seem to be needed for this to work.

- All of the superpartners are expected to be unstable particles, decaying into lighter superpartners—except for the lightest superpartner (LSP), which has no lighter ones to decay into and is thus stable. Thus supersymmetry introduces a new stable particle into the universe, joining photons, electrons, neutrinos, and protons. The light we see from stars is composed of photons. Protons and electrons form the stars and planets. Neutrinos, and the LSP (if it exists), will be forms of matter that are present throughout the universe. Since neutrinos feel only the weak and gravitational forces, not the electromagnetic or strong ones, they will not participate in forming stars. They will be "dark matter." Supersymmetry predicts that there is also dark matter composed of the LSP. Right after the big bang there were about the same number of each kind of particle. Most particles decayed into lighter ones, and some annihilated into others as the universe expanded and cooled. Since we have a theory of how they all interacted, we can calculate how many are left now. Even though they did not coalesce into stars that produce photons that we can see, their presence can be detected by their gravitational attraction for what we do see, if there are enough of them—their presence modifies how stars move in galaxies and how galaxies move relative to one another. It was realized in the early 1980s that supersymmetry predicted that there should be considerably more LSP dark matter than even the matter in stars. Astronomers had indeed already observed that the universe did have considerable dark matter, because stars and galaxies did not move through the universe as they would if the only matter was what we could see, but at that time it was not known experimentally if the dark matter could be nonluminous forms of ordinary matter (e.g., interstellar dust) or if a previously unknown kind of matter must exist. We will examine the dark matter and how to study it in the laboratory in Chapter 6.

- The LEP collider at CERN and the SLC collider at Stanford were constructed during the 1980s in order to test the Standard Model and to search for new physics that would strengthen the foundations of the Standard Model. Before LEP and SLC operated, it was possible to predict the kinds of results that they should find in a supersymmetric world. Either superpartners would actually be produced and detected or else the effects of supersymmetry on the quantities measured were expected to be very small, so observables should have essentially the values that the Standard Model predicts for them. Nonsupersymmetric approaches to an explanation of the Higgs physics generally predicted larger effects. After a decade of accurate measurements, no superpartners were produced, and the results did confirm the prediction that there were no sizable deviations from the Standard Model expectations. (If you have sufficient confidence in indirect and aesthetic arguments, you might even conclude that the need for some extension of the Standard Model to explain the Higgs physics, plus the absence of any deviations from the Standard Model expectations for observables in the LEP and SLC data, together confirm that supersymmetry must be a part of the description of the world. But most physicists insist on verification that is direct.)

• We have already seen that supersymmetry provides two (consistent) ways that suggest that the description of the electromagnetic, weak, and strong forces will be unified with gravity. One was the way they became of the same strength at very small distances near the Planck scale. The second was that supersymmetry as a symmetry throughout space and time necessarily had a connection to the theoretical description of gravity (more on this below). While the connection of supersymmetry to gravity is not yet fully understood, it is very encouraging that the supersymmetric Standard Model is related to the theory of gravity, while the Standard Model is not.

• In Chapter 8 we will examine the connections of M/string theory and supersymmetry. M/string theory seems to require supersymmetry as an essential part of the theory, though we do not yet know if this is truly a necessary condition. In order to learn how M/string theories work, and whether their solutions might really describe our world, we must learn how to solve the equations of the theory. Finding solutions has turned out to be much easier (and therefore sometimes possible) by using the powerful constraints that have to be satisfied when the theory is supersymmetric.

• Supersymmetry has led to new approaches to solving or explaining a number of other important problems that had no possible solution in the Standard Model. Here some will just be listed to give a sense of the opportunities, without detailed explanations or jargon; later we will examine several of them. Fundamental questions for which supersymmetry provides new ideas and methods include understanding how the universe got to be mainly matter and not antimatter, whether and how protons decay, why the universe is the age and size it is, rare decays of quarks and leptons, and more.

• Finally, probably the most important consequence if supersymmetry is part of our description of nature is that it provides a window to look at the Planck scale from our world that is so distant from the Planck scale. We have seen that there are several reasons to expect that the final theory will be naturally formulated at the Planck scale. But we cannot ever hope to do actual experiments at the Planck scale—it is just too small. If supersymmetry is part of our description of nature, the implication is that we can write the predictions of a candidate for the final theory at the Planck scale and then calculate what should be observed in experiments we can do at laboratories and colliders—for neutrino masses, for proton decay, and much more. Similarly, we can measure parameters that are part of the description of the theory in experiments and then calculate the values they have at the Planck scale (many of them have values that depend on the distance scale, which is common in quantum theory). Most likely the latter procedure will be the actual one, with experimental input needed before it is possible to formulate detailed candidates for the final theory: we can hope someone is smart enough to guess the final theory, but if not we can get there anyhow by building up knowledge about the form the theory must take, using experiment and theory together as physicists have traditionally done so well. Even if someone does guess the final theory, that person won't be sure, and no one will believe it, unless there is experimental input of the kind that supersymmetry can provide. If the world is not supersymmetric, we do not know of any way to relate physics at the Planck scale to physics at our scale. If the world is supersymmetric, we can expect to be able both to formulate the final theory and to test it.

The above arguments were of two kinds. Some provided explanations for actual phenomena, or predicted observations, while others were about our opportunities to make sense of the world at the deepest levels. The first kind provides real evidence that the world is indeed supersymmetric—indirect but still real and significant evidence. The second kind, of course, does not imply that supersymmetry is real; the mere fact that it will help us solve M/string theories or provide a window on the Planck scale does not mean nature is that way. But both kinds of reasons contribute to the very strong and widespread interest in supersymmetry among physicists. And it is worth repeating that supersymmetry was not invented for any of these reasons—all the reasons emerged as its implications were studied. Indeed, the results described above provide impressive examples of emergent properties of a basic theory, properties implied in a sense by the theory but not apparent until suggested by data or after much study.

Supersymmetry has the consequences it does because the requirement that the theory is unchanged when bosons and fermions are interchanged can be satisfied only if the theory (the set of equations it comprises) is constrained to have a certain form. It may be helpful to mention a couple of historical examples of properties that emerged as theories became more constrained. Maxwell took several equations that people had developed over decades to explain electrical and magnetic phenomena and tried to combine them into one consistent set. He found that to do that he had to add a term to one equation. The new set of equations then turned out to have a new solution that described light as an electromagnetic wave: suddenly the behavior of light was incorporated into the electromagnetic theory, and electromagnetic waves we now use for communication were unexpectedly predicted to exist. Another example comes from special relativity. Einstein found that in order to have the same laws of electromagnetism hold in all systems shifted in position or speed relative to one another, it was unavoidable that space and time got tied together. In any mathematical science, new constraints on the basic equations often imply major new physical phenomena, just as we have seen happen in the case of supersymmetry.

In Chapter 2 we saw that an important aspect of the Standard Model was the invariance of the theory under the interchange of certain particles—the electron and the electron neutrino, the up and down quarks, and so on. Some of the particles required for that invariance to be valid already were known to exist when the Standard Model was developed, but others were later predicted to exist and then they were found. We also saw that having the theory be consistent both with quantum theory and special relativity required the existence of antiparticles. None of the antiparticles had been observed at the time they were predicted, and because physicists had learned what to look for, all have been observed since then. For supersymmetry to be valid, it would again be necessary for previously unknown particles to exist that are the same as the particles of the Standard Model, except that bosons ↔ fermions. Since bosons and fermions have different spins, the partners must differ in their spins. There must be another particle just like a photon, with no electric charge or weak charge or strong charge, but with spin one-half instead of spin one. There must be another particle just like the electron, with the same electric charge and weak charge as the electron, no strong charge, but with spin zero instead of spin one-half. An examination of the particles we know shows that none of them have the properties to be the needed partners. The situation is analogous to what happened with antiparticles, whereby all the predicted new particles had yet to be found. We call the new set of particles "superpartners," or "sparticles." The sparticles form smatter.

As our understanding of nature progressed through the past century we saw the number of "electrons" grow, in a sense. From 1895 till the 1920s, there was thought to be only a single electron. Then first the data about energy levels of atoms, and later Dirac's unification of quantum theory and special relativity, led to assigning half a unit of spin to the electron. In quantum theory that meant the electron spin could point up or down with respect to any arbitrary direction, and accounted for the observed extra energy levels of atoms. It is as if one should think of two electrons, one with spin up and one with spin down. Then antiparticles were predicted and found. Again, it is as if electrons had to come in four kinds, particle or antiparticle each with spin up or spin down. With the validation of the Standard Model in the 1970s, we learned that electrons and neutrinos were really different projections of the same basic particle that could turn into each other by emission or absorption of a W boson, with analogous processes for antielectrons and antineutrinos—so in a sense we think of one object that can exist with spin up or down, and as electron or antielectron, neutrino or antineutrino. With supersymmetry we go one more step. Each of these states of the "electron" is predicted to have a partner that differs only in its spin being zero rather than one-half. Finally there are 16 electrons.

This might seem to be a proliferation of particles. But if the theory says that given any one of them all the others must exist, then the apparent proliferation follows from a conceptually simple structure. Just as a simple structure, based on the electron and the up and down quarks and their interactions, underlies the great complexity of our world, so there is a simple conceptual structure with a small set of Standard Model particles plus their antiparticles and Standard Model partners and superpartners.

There is a simple notation and terminology for superpartners. The superpartner of every particle is written as the partner with a tilde ($\sim$) over it: $e \rightarrow \tilde{e}$, γ (photon) $\rightarrow \tilde{\gamma}$, and so on. If the Standard Model particle is a fermion, the name of the partner is the fermion name with an "s" in front—sfermion, selectron, squark, up squark, top squark (or, equivalently, stop), sneutrino, and so on. If the Standard Model particle is a boson, the name of the partner is the boson name with an "ino" suffix—photino, gravitino, Wino, higgsino, and so on. Since all of the regularities of the Standard Model hold for the supersymmetric Standard Model too, keeping the names helps in keeping track of processes and predictions. As many of us have said, we have a new slanguage spoken by sphysicists.

It should be emphasized that supersymmetry is the *idea* that the laws of nature are unchanged if fermions $\leftrightarrow$ bosons. It is not that an electron becomes a selectron in the equations that constitute the basic theory but, rather, that the equations should contain symbols representing both electrons and selectrons, and the equations are unchanged if those symbols are interchanged. The existence (or not) of the sparticles themselves is the most dramatic prediction to test that idea—they are the anticipated smoking gun. There are several other tests that are less explicit. Some indirect tests based on the implications of supersymmetry have already been positive for supersymmetry, helping convince many physicists that it is indeed correct; those are the ones listed earlier in the chapter, and we will examine some in detail in later chapters. Supersymmetry requires not only new (to us) kinds of particles but also new interactions involving the new particles. The theory implies that the new interactions have related strengths. That's an important prediction, and will be an important test of the theory if some of the selectrons and photinos and other sparticles are observed.

SUPERSYMMETRY AS A SPACE-TIME SYMMETRY: SUPERSPACE

We have seen that there are two important ways to think of supersymmetry. They are consistent, of course, but often one or the other is most helpful to achieve a given goal. The first and most common way is to think of it as a symmetry of the laws of nature under the interchange of bosons and fermions. A sparticle is predicted to exist for each of the quarks and leptons and bosons of the Standard Model. The experimental approach to supersymmetry and most practical calculations are carried out in this framework. A second approach is to think of supersymmetry as an effective theory that opens a window on the Planck scale, so that we can formulate and test the final theory; this shows us how important supersymmetry is in our search to understand our universe. There is a third way. Sometimes it is fruitful to think of supersymmetry as a space-time symmetry, but in an extended space-time, called "superspace." This approach is often most helpful in uncovering the mathematical properties of the supersymmetry theory. The remainder of this section is a little more technical than subsequent chapters and is not needed to follow the rest of the book.

Think about an isolated particle or object, one on which no forces act. Assume it's moving, with some energy and momentum. If there were forces they could change its energy and momentum, so with no forces we expect the energy and momentum to be unchanged. We say the energy and momentum are "conserved." We are making assumptions when we assume that the energy and momentum are conserved in the absence of forces. We expect space and time to be neutral, so the object can move through them without feeling effects. If no other objects are around, we don't think it matters whether our object is here or a few miles away, or rotated some amount, because we think space and time don't affect objects in them. Whenever something is invariant under changes, we speak of a "symmetry." Whenever some quantity is conserved, there is an associated symmetry, and vice versa. Just as we can change our position in space or time and find symmetries, we can change our position in superspace and find an associated symmetry—supersymmetry.

Symmetries in physics can be of two kinds. The examples we just considered are "geometrical" or space-time symmetries. In Chapter 2 we saw that the Standard Model has some symmetries that can be called "internal" ones, the interchange of the up quark and the down quark and so on. A very important question is whether there could ever be a symmetry that mixes geometrical and internal symmetries. For example, we can be pretty sure that moving an up quark a few centimeters won't change it into a down quark, but maybe there is some more subtle geometrical change that could? It turns out that for our standard ideas of space-time the naive approach is right, and geometrical and internal symmetries cannot be mixed. It can be proved that the space-time symmetries we already know about are the only possible ones. However, if we allow for new "fermionic" dimensions, then it turns out that one more symmetry can exist—and it is supersymmetry. Another attractive feature of supersymmetry is that it is the last possible space-time symmetry that nature could have, the only mathematically possible symmetry that is not already known to be part of nature.

Supersymmetry is the idea that the basic laws are invariant under interchanging fermions and bosons. We just saw that invariances lead to symmetries. In our usual space-time, bosons and fermions behave differently. It turns out that we can attach to our normal space-time another four "dimensions" that will allow us to incorporate differences between fermions and bosons into the structure of the "space," defining a "superspace." Superspace is a geometrical structure in which fermions and bosons can be treated fully symmetrically.

The extra superspace dimensions are not like our physical dimensions. They have no size at all. Nor are they like the extra small spatial dimensions of M/string theory. In the superspace we can think of bosons and fermions as two different projections of a single object, similarly to the way an electron and its neutrino can be thought of as two different projections of a single object in an internal space. Superspace is an intrinsically quantum theoretical structure, because the whole idea of fermions is meaningful only in a world described by quantum theory. One can think of superspace as adding fermionic dimensions to our usual bosonic dimensions. If the world really exists in superspace, it affects observables—it is a testable idea. The simplest and most dramatic test is that the superpartners must exist. If superpartners are found, we can interpret the results as letting us learn experimentally about the properties of superspace.

One can turn the history around. Suppose from the beginning quantum theory had been formulated in superspace. Then immediately it would have predicted that all particles would exist in both a bosonic and a fermionic form. Supersymmetry, formulated in superspace, requires that fermions exist—in a sense it explains their existence.

Einstein's general relativity is a geometrical theory of gravity, which views the gravitational force as an effect of the distortion of space-time by masses (e.g., planets). Once superspace was formulated, people immediately thought of using it as the basis of a generalized geometrical theory of gravity, "supergravity." Supergravity incorporates general relativity, and extends it. The graviton that mediates gravity is predicted to have a superpartner, the gravitino. The connection of supersymmetry to gravity encourages people to think that the unification of gravity with the Standard Model forces will include supersymmetry. Interestingly, the geometrical structure of supergravity takes a simple and elegant form if the space-time is extended to eleven space-time dimensions (with associated superspace coordinates). As it happens, that is also the largest space-time dimension in which one can write a consistent theory of gravity.

HIDDEN OR "BROKEN" SUPERSYMMETRY

So far this book has argued that there is good reason to expect that the laws of nature are supersymmetric. In fact, the situation is more subtle. We know that nature is not exactly supersymmetric for two kinds of reasons. First, if the world had a selectron with every property identical to that of the electron except its spin, we would already have observed it in experiments. That's true for a number of other sparticles too. Second, even if the relevant experiments were not possible, we can deduce that the world would be entirely different if selectrons with the same mass as the electron existed. In fact, there would probably be no life in the world. Indeed, if bosonic electrons existed they would all fall into the lowest energy level of an atom, shielding the electric charge of the nucleus, and there would be no valence electrons to bind atoms into molecules. This reason should not be taken as an excuse for supersymmetry to be broken, but it illustrates how indirect analyses can be helpful in understanding how the world works.

In physics, symmetries can be powerful guides to how the systems behave. Some symmetries are exact ones, while many are not. Even when they are not exact they still can be very helpful in understanding the behavior of the system. As a very simple example, think of a child's spinning top. It has a cylindrical symmetry about the axis that it spins around. Suppose it were poorly made, somewhat bumpy. It would still spin. Maybe it would wobble a bit if it were a little heavier on one side. It may spin a little less long and be more likely to end up lying on one side than another, but it is still recognizably and basically a top even though its cylindrical symmetry is not quite right. When a symmetry is only partly valid or perfect, physicists speak of it as a broken or hidden symmetry. An example closer to our present interests is the symmetry between particle and antiparticle found by Dirac from which he deduced the existence of antiparticles. That symmetry was called "charge conjugation invariance" since Dirac's equation did not change when the sign of the charge was reversed. It is a valid symmetry for the particles alone, and also valid for the ways the particles interact via the electromagnetic and strong forces, but it turns out not to be valid for interactions via the weak force. The important point here is that despite the partially broken symmetry, all the predicted particles (i.e., the antiparticles) have been found to exist. The supersymmetry case is expected to be somewhat similar to the antiparticle one. All of the superpartners should exist, but they can have masses different from their particle partners. The interactions will be similar, but they can differ because the masses are different, and for other associated reasons. If the superpartners are heavier, they might be too heavy to have been observed in experiments so far.

The issue of hidden supersymmetry illustrates well how physics progresses. In the 1960s, there was no theory of the forces or basic particles at all. Then the Standard Model emerged. It provided a very good description of the particles and interactions. Its remaining problem was to determine the form the Higgs physics takes so that the particle masses could be incorporated into the theory, and to understand from what underlying physics the Higgs physics itself arose. Supersymmetry in turn can provide the answer to the Standard Model Higgs physics problem; that is, it explains how the Higgs physics arose. The supersymmetry answer actually depends on the way that supersymmetry is hidden or broken, and it ties the supersymmetry breaking to the Higgs physics of the Standard Model (as explained below). We do not yet understand how the supersymmetry is broken—that should be explained by the next level of effective theory; probably the supersymmetry breaking won't be explained within the supersymmetic Standard Model itself. But supersymmetry is a sufficiently comprehensive theory that it allows us to know the form the broken supersymmetry theory must take, even though we do not yet understand the mechanism that generates that form—very much like the way we were able to incorporate the masses using the Higgs physics even though we did not yet understand how the Higgs physics arose. A short-sighted view might claim that we had just traded one problem for another, but that is really not so: on the one hand, the new extended supersymmetric Standard Model incorporates a variety of phenomena that were mysterious before it emerged; and, on the other, the supersymmetric Standard Model is one effective theory closer to the final theory. If nature is indeed supersymmetric, explaining the origin of supersymmetry breaking will become the central problem of supersymmetry theory and experiment.

The manner in which supersymmetry explains the Higgs physics is elegant and has important consequences for how we expect to test supersymmetry experimentally. It is rather technical. There are three parts to the Higgs physics of the Standard Model. First, the Higgs field must exist. Second, it must interact a certain way with the other particles—that is called the Higgs mechanism. Third, the quanta of the Higgs field, one or more Higgs bosons, must exist—that is required once the Higgs field exists. The supersymmetric Standard Model automatically contains fields like Higgs fields, so it makes their existence more natural. Supersymmetry also provides the interaction, the Higgs mechanism. The crucial signal that the Higgs mechanism is working is that a parameter of the theory that naively should be interpretable as the square of a mass (let's call it M^2 to give it a name) and therefore a positive number, is in fact negative. While that seems at first to be a difficulty, it turns out on deeper examination to do what is needed to give the other particles a mass. So one can tell if the Higgs mechanism is working in any theory by checking to see if the quantity called M^2 is negative. In the supersymmetric Standard Model M^2 is one of the supersymmetry parameters, and it has a positive value at the Planck scale so all the quarks and leptons and W and Z are massless at that scale. But quantities such as M^2 aren't constant; rather, they vary as the theory is applied at different distance scales. We need to know M^2 at the scale of the weak interactions and larger scales, where we live and where our experiments show that the quarks and leptons and W and Z have mass. In the supersymmetric Standard Model we can calculate how the value of M^2 changes as it goes from the Planck scale to the weak scale. The remarkable result is that if one condition holds, M^2 decreases to zero and becomes negative at larger distances, inducing the Higgs mechanism!

That condition is that one of the quarks has to be rather heavy, heavier than a W boson. When the way supersymmetry could explain the Higgs mechanism was first pointed out in the early 1980s, the heaviest known quark was far lighter than the W boson, but the mass of the top quark was not known. So supersymmetry predicted that the top quark would be far heavier than the naive estimates. When the top mass was finally measured in the 1990s it was indeed about twice the W mass, comfortably satisfying the supersymmetry prediction. If history had been a little different and supersymmetry had been developed and understood before the Standard Model, the Higgs mechanism would not seem mysterious at all, but just a new and unexpected consequence of supersymmetry.

The way supersymmetry explains the Higgs mechanism leads to an important result: it is the only way we know to relate the masses of the superpartners to known masses, so we can estimate how large the superpartner masses are. The Higgs mechanism leads to a description of the masses of the Standard Model particles in terms of the masses of the superpartners. So one obtains an equation with a known Standard Model particle mass (W or Z mass) on one side and unknown superpartner masses on the other. For any equation like that, one would not trust the result if the quantities on one side were much larger than those on the other, because any measurable quantity in a physics theory can be estimated only to some accuracy; there are always experimental and approximation errors involved. Therefore, the supersymmetric Standard Model explanation of the Higgs mechanism would not make sense unless some superpartner masses were not much larger than the Standard Model masses they explain. That gives us an estimate of the masses we should expect the superpartners to have as we search for them. Such estimates are only approximate, but luckily the expected masses are small enough that they imply the superpartners should be detected soon. The next chapter introduces the experimental search, followed by a description of the LHC and the detectors and analysis.

5
....

Finding and Studying Supersymmetry

On July 4, 2012, the Director General of CERN and the two general-purpose detectors, ATLAS and CMS, announced the discovery of a new particle, a Higgs-like boson. The main goals of collider experiments now are studying the properties of the Higgs-like boson and finding superpartners (or, if somehow we are on the wrong track in spite of the indirect evidence and strong theoretical arguments, showing that the superpartners do not exist with the expected masses). The exciting discovery of the Higgs boson both completes the Standard Model and provides clues about how to extend the Standard Model and strengthen its foundation. The meaning and implications of the Higgs boson discovery are the focus of Chapter 7. This chapter describes how to recognize a superpartner mainly at the Large Hadron Collider, and Chapter 6 looks at the special properties of the lightest superpartner that make it a very good candidate for the cold dark matter of the universe, as well as at special ways to detect it in actual experiments.

DETECTORS AND COLLIDERS

How can we get experimental evidence about aspects of the world that are too small or too far away for us to see directly? That question would be a good unifying theme for a history of scientific discovery. We can start by asking how we see anything in our world. Photons bounce off objects and then enter our eyes, forming images after significant processing by our brains. People first looked deeper than we can see with the unaided eye by making glass lenses that magnified images. Optical microscopes and new techniques will let us see things thousands of times smaller than the unaided eye can see, down to atomic sizes (10^{-8} cm). That is still ten orders of magnitude larger than we need to study to look for Higgs bosons and superpartners. In the natural world the most energetic particles we can control to study something come from the decay of nuclei; they have an energy about a million times greater than that of visible photons, so in effect they can help us "see" something a million times smaller. To probe even more deeply, physicists had to invent accelerators that would increase the energy of particles. That happened in the 1930s, and accelerators have been improving since.

Of course, we do not see the particles the same way we see photons. It is necessary to add an intermediate stage, a detector that collects the information about the behavior of the particles and then presents an image for us. Often it is an image we see, though sometimes even apparent images have more stages between the object and ourselves; for example, present-day telescopes often really show us digitized interpretations of light outside the visible range. The first experiments used beams of particles from nuclear decay to observe how they bounced off atoms, by surrounding the target with screens that would fluoresce when hit by a particle. Then people could observe the light from the fluorescing screens. The people were one step removed in the observation process. The information could have been recorded on film and studied later, adding more stages. Today, often several stages are present before the information gets studied by a person. A camera is a detector—it sees by recording the effects of photons. The detectors that are used in particle physics have to also "see" electrons, muons, and other particles in order to reconstruct the information from the collisions, so they are much bigger and more complicated than cameras.

At a detector, a collision occurs. As much information as possible is extracted from the energies and directions of the emerging particles. From the information they are able to extract, the experimenters have to figure out quickly if the event is one expected in the Standard Model, or if it is "new physics." At the LHC there are two very large multipurpose collider detectors. The probability of a new-physics event occurring is the same for each of them. The detectors were designed to have somewhat different properties, so one or the other might be more or less sensitive to a particular kind of new physics. A heroic effort is needed to design, construct, and use a detector to experimentally probe nature to a millionth of a billionth (10^{-15}) of what we can see with our eyes. Though they are the largest scientific devices ever built, the detectors are the smallest "microscopes" that would allow us to observe the superpartners; in effect, they are "superscopes."

Einstein taught us that energy and mass were in principle interchangeable, and eventually people realized that this provided a new way to get information about particles. If particles could be made very energetic, and then be made to collide with other particles, some of their energy could be converted into the creation of previously unknown particles. Sometimes we'll call them new particles, even though they are new only to us, not to nature. The first new particles were detected in the 1930s, when people set up detectors to observe cosmic rays that hit a nucleus in or near the detector and converted most of their energy into making particles. (Cosmic rays are particles impinging on the earth from outer space; sometimes they are very energetic, the most energetic ones mostly originating from exploding supernovas.) Later, physicists learned to accelerate electrons and protons to higher energies with electric fields and, eventually, to collide them so that even more energy was available: when a bat and a ball collide, the ball will go a lot further (i.e., get a lot more energy) than if it had hit a stationary bat. In practice, as we will see in more detail below, only processes allowed by the Standard Model forces and the rules of quantum theory can generate new particles, so the experiments have to be planned carefully. When particles are produced in a collision, they are not particles that were somehow inside the colliding particles. They are really produced by converting the collision energy into mass, the mass of other particles. A particle collider is a device for using $E = mc^2$ to make new particles. Which particles will be produced is partly determined by their mass—the lighter they are the easier it is to produce them, other things being equal—and also by the probabilities calculated from the forces and rules. In this chapter we focus on the ways sparticles can be produced and detected in experiments, thus hopefully confirming that supersymmetry is part of our description of nature if the superpartners are found.

Suppose superpartners are produced in an experiment. All but the lightest are unstable and typically decay so quickly that they do not hang around and let us study them directly. Fortunately they usually decay in unique and identifiable ways, so it is possible to deduce their presence; we'll examine how later in this chapter. The lightest superpartner may constitute much of the dark matter of the universe; if so, there are ways to observe it in laboratory experiments. That is the subject of the next chapter. In Chapter 4 we examined some indirect arguments for supersymmetry. However convincing the indirect evidence may be, supersymmetry will not be accepted as an essential part of our description of nature, and we will not be able to use it as a window on the Planck-scale physics, until some superpartners are observed and studied at colliders.

Whether a particular collider can produce superpartners in a collision depends on the energy and intensity of the colliding beams of particles, and on the masses of the superpartners. More energy is needed to produce heavier superpartners. More intensity may be needed to produce enough superpartners to detect a signal. The effective intensity of a beam depends on how many particles are in the beam, how closely they are packed together, and other considerations. For our purposes we can just think of it naively: colliding more intense beams is more likely to produce new particles. Once they begin to appear, more can be produced and they can be studied.

Because we have to probe ever more deeply into nature to gain new insights, the "microscopes" we have to use become larger and more expensive, take longer to build, and require larger groups of scientists. Sometimes people lament the loss of the golden age when experiments could be done in a few weeks by one or two scientists. That era is long gone at the frontier of particle physics. At the beginning of the twentieth century, Marie Curie stirred huge, hot, smelly, dangerous pots filled with pitchblende to extract a tiny bit of radium, for about as many years as it took to build LHC.

In a collision in which energy is converted into massive particles, the number of new ones produced could be one, or two, or more, depending on the properties of the particles. Because the total amount of charge in a process cannot change, and the new particles often carry some charge whereas the initial pure energy carries no charge of any kind, often two new particles are produced (with opposite signs of the charges, so the net charge produced is zero). Thus typically only particles of mass about half of the total available energy or less could be produced.

The available energy is not the only consideration. The rate for producing each new particle depends on the particle's mass and on details of how it interacts. Suppose the collider had beams of sufficient intensity to produce one new particle event per year on average. In a given year there might or might not be an event, and even if there were, with only one event we might not be convinced that something new had really occurred, because detector malfunctions or an event coincident with a cosmic-ray collision or some odd thing could perhaps fake an event. Depending on how common fakes could be, we might not be confident of a signal until ten, or twenty, or some larger number of events had occurred.

RECOGNIZING SUPERPARTNERS

Recognizing the presence of superpartners may not be easy. It's not that we don't know how they will appear. Supersymmetry is a very well-defined and powerful theory—in fact, the main thing that we don't know about it is whether it actually describes nature. For practical purposes we also need to know the masses of the superpartners to predict their decay patterns; but given those, everything else is determined. If we could calculate or guess the masses, then we could calculate the number of superpartners that would be produced at the LHC and how they would appear in the detectors. Since we don't yet know the masses, we simply do calculations for a variety of assumptions about the masses. Then we examine the results to see what typical patterns of final particles are likely to appear in the detectors, and plan to look for those.

Several kinds of problems arise. One is practical. You can measure more with a higher-quality detector. But that costs more, and the funding for detectors is limited. Sometimes the technology doesn't yet exist to have the detector do what is needed. The experimenters have done an amazingly good job considering all the constraints. Sometimes different people collaborating on the detector have different physics goals—having the detector be more sensitive to one kind of physics may decrease its sensitivity to another if there is a fixed total funding. If signals for superpartners seem to be in the data, most likely it will be possible to improve the detectors to confirm the signals and identify their implications. Having two general-purpose detectors provides powerful checks on each.

The second problem is different. Standard Model processes lead to the production of heavy particles too, Ws and Zs and top quarks. Sometimes these can mimic the new signatures expected for supersymmetry. These events are called "backgrounds" for the new physics. Like the images from cameras, those from detectors are often a bit fuzzy. With perfect images it would be possible to tell which was which, but in practice it is harder.

A third problem is intrinsic to supersymmetry. In any collider process, two superpartners will be produced. Each either is the lightest superpartner or will decay into the lightest superpartner and other particles. (In some cases the decays can occur after the superpartner has exited the detector, but we'll ignore complications like that.) So there will be two lightest superpartners in every event. It turns out that the lightest superpartners interact very weakly with the ordinary matter that detectors are made of, so they are not directly detected. Instead they can each carry off energy in any direction. Always before when previously unknown particles were produced at colliders it has been possible to find a unique directly visible signature of the new particles, with some of the decay products or combinations of them having a unique energy that stood out from any possible background. In the supersymmetry case that can never happen, because of the two missing lightest superpartners, so it will be much harder to convince anyone who is not an expert that a real signal has been observed. On the other hand, since extra energy is carried away compared to typical nonsupersymmetric processes, one of the main ways to search for supersymmetry is to build detectors that are very good at detecting all of the energy of the Standard Model particles in an event and summing it up. This possible "missing energy" effect will be the first thing searched for at every new collider and collider upgrade.

For completeness I mention that the expectation that the lightest superpartner is stable may not hold. The effective supersymmetry theory allows the lightest superpartner to decay to Standard Model particles. M/string theory addresses the question of whether the LSP stability can be derived, and in some versions of the compactified M/string theories it can, but not in all. If the LSP was not stable, it would not form part of the dark matter. I will not go further into this variation, but collider searches allowing for this alternative will be done, at least until there is evidence for LSP dark matter.

There is another fascinating issue that has a lot to do with making detectors complex and difficult to build if the goal of the detector is to "see" new physics. Because collisions have been studied for over half a century, almost all of the collisions produce familiar and well-described results. Only a few events—fewer than one in 10 billion—are expected to contain superpartners! If the detector records all the events, it becomes effectively impossible to pick out the superpartner ones; for example, if one could examine one event a second to decide if it contained superpartners, it would probably take more than a thousand years until the first one was found. So the detector must be designed to make decisions in billionths of a second about whether to retain an event as one with a candidate superpartner. If it does that wrong, it will miss such events even if they are there.

Once a possible signal is seen (by "possible signal" I mean a statistically significant set of events not expected from the Standard Model), we do understand a lot about how to become confident that it really is a signal. First, people carefully calculate the Standard Model backgrounds and demonstrate one by one that none of them can be misinterpreted as such a signal. Second, it is necessary to demonstrate that the observed production rates and different decay modes and masses of the superpartners are consistent with what is expected. That is a very strong constraint. The apparent production rate for a proposed signal for superpartners could be much larger than the expected one, in which case we would know it was not a real signal; it would be expected to be either a technical error in the analysis or a statistical fluctuation that would go away as more data were gathered. Or the "observed" decay might be a minor one, implying that the major decay must also be present, but the major decay was not found to be in the data, again implying the signal was not real. If a signal passes all such tests it can be taken very seriously. Finally, once one sparticle signal is found, others have to be nearby. They can be looked for in different signatures. This part of the study is less certain initially because what is expected can depend on the unknown masses; but once some masses are measured, additional ones can be estimated. These kinds of tests are possible here because supersymmetry is such a well-defined and well-understood theory.

Sparticles and Their Personalities, Backgrounds, and Signatures

Those of us who think about sparticles daily form images of their behavior. We like some more than others. Selectrons and smuons, interacting with photons and Zs, decay into electrons or muons plus the lightest superpartner and have no interactions via the strong force. But production of a W pair can look a lot like selectrons or smuons unless one looks very closely. Winos and charged higgsinos do a complicated thing: when the Higgs mechanism operates they can't be told apart, so what we see will be a mixture, called "charginos." Similarly, photinos and zinos and neutral higgsinos can't be told apart, and we will observe mixtures called "neutralinos." Gluinos have the strongest interactions of any superpartners so, if they are not too heavy, more of them will be produced at the LHC than any other sparticle.

Several arguments imply that some sparticles are within the reach of the LHC. One of the most appealing is based on the explanation that supersymmetry gives for the Higgs mechanism of the Standard Model, as described in Chapter 7. Basically the qualitative argument is that since supersymmetry provides the Higgs mechanism that accounts for the masses of W and Z, some sparticle masses cannot be much heavier than the W and Z masses themselves. When this argument is put in a technical form, it implies that gluinos and probably charginos and neutralinos should be within the LHC reach. There are some arguments, both theoretical and phenomenological, suggesting that squarks and sleptons will be too massive to produce at the LHC.

Suppose superpartners are indeed detected at the LHC. What then? There will be great excitement, because we will have dramatically increased our understanding of how nature works, and of the path to follow to study it, and most importantly because we have further opened the window to the ultimate, or final, theory. Just knowing that nature is indeed described by supersymmetry would tell us that we are working in the right direction to make more progress. Supersymmetry may be the last stage on the path to the final theory that can be studied experimentally.

Then the fascinating work begins. The properties of the superpartners can tell us an immense amount about how the final theory works. In trying to formulate the final theory we can imagine many alternative approaches—each approach will lead to superpartners with different masses and interactions. So it will be very important to take detailed data to deduce the needed properties. It will be a challenging task, but an exciting one.

FUTURE COLLIDERS?

While the first signals of supersymmetry should be found soon if present ideas are basically right, perhaps the LHC will not be sufficient to fully untangle the information needed to tell us how supersymmetry is broken, or to point strongly toward a particular way to extend the Standard Model.

Since colliders are always at the frontier of what is known, existing techniques and technologies must always be extended in order to design and construct new ones and their detectors. It necessarily takes considerable time to be sure they will work as intended and reliably for years on end, and to find relatively inexpensive ways to construct them. The construction of new facilities itself takes years. So, unfortunately, it is unlikely that new devices will suddenly emerge and give us unexpected new ways to produce new particles and study them.

CAN WE DO THE EXPERIMENTS WE NEED TO DO?

Will we need colliders that have even larger energy or intensity? Can such colliders be built? One sort of limit could be financial. Facilities that can probe nature more deeply than ever before, so we can continue the quest to understand the physical universe, have to be larger and more complex than previous ones. They necessarily involve new techniques, since the older facilities would have been extended if that were possible and likely to provide new knowledge. So they are expensive. Only governments or the largest foundations or a few wealthy people could build them.

In fact, society always comes out ahead even from a purely economic perspective when it builds such facilities, because new developments lead to "spinoffs" that in turn lead to startup companies and even to multibillion-dollar industries. A recent example is the World Wide Web, introduced by particle physicists to handle new challenges involving analysis programs for detectors as well as unprecedented amounts of data. There are many more examples. But the political leaders who decide on the funding may not understand that it is an investment. Even worse, the returns on the investment are not so likely to occur during the term of office of the relevant politicians. Society should be funding every scientifically and technically justified experiment. Experience shows that it is not a zero-sum battle—every part of society wins if these projects go ahead.

The financial barrier is partly caused by how governments and institutions function. The original CERN charter that brought the main European countries together in the 1950s to set up a scientific laboratory contained a wonderfully wise or lucky feature: the contributions from each of the member-countries were line items in the country's budgets, indexed to inflation. Line items are seldom discussed in any detail or voted on. In the United States and most other countries, the budgets for large projects are instead reconsidered yearly even though a project can take a decade to build. For political reasons their budgets can rise or fall annually. Then equipment cannot be ordered in a timely way, and sometimes needed talented people are lost. Such outcomes were largely responsible for the failure of the United States' Superconducting SuperCollider. That the CERN budget is effectively stable has been crucial to the success of CERN. CERN's current plan is to upgrade the LHC's luminosity and energy in 2014. Depending on what new physics is discovered and where it points, one or another kind of future collider will be planned.

Another limit could come from a lack of appropriate people. Significant numbers of talented and committed people are needed to build a forefront facility and make it work. It's not possible to start from zero. At any particular time the people are there because they have worked on the previous facilities that have been stretched as far as they go, and because they can see how to answer major scientific questions with new facilities. If progress is blocked, the people have to have jobs—so they disperse. It is not possible to reassemble them. Society loses not only because the new developments do not occur but also because talented people are forced out of research. In cases where the world does not support the possible and relevant facilities, the associated research fields will die. Some of the best researchers will work elsewhere, and some will leave research altogether. The remaining ones are less likely to do the job right.

Whether we will need more powerful colliders depends on whether we find the sparticles and the other Higgs bosons in the coming years, and whether there are surprises in what is found at the LHC. One can imagine a world in which all of the evidence about the final theory that could ever be found at colliders can be found at the LHC, but we won't know if the world is that way until we get there. One can also imagine worlds where that will not happen. I strongly emphasize that the possibility that one can imagine such a world is a very strong statement—one that could never have been made before in history—because now we have the Standard Model and we have supersymmetric approaches that can provide not only a window to the Planck scale and the final theory but also help with the study of the unanswered questions about the universe (more on this later). In addition to the new particles and interactions that can be studied at colliders, we need information about which interactions are invariant under time reversal and particle-antiparticle interchange as well as about dark matter, the stability of the proton, neutrino masses, and more. The experiments that can tell us about these issues are also challenging, but possible.

Even if we need more energetic or intense colliders, we may not be able to have them. The only way to increase the energies of the particles is with electric fields, and there are practical limits to how strong an electric field we can make in a situation that could accelerate particles. The higher energies imply that the length of the acceleration part of the collider must be longer if this part is a linear one—eventually so long that it becomes too expensive or impractical. If magnets are used to curve the path of the particles into a circular one, constraints on how energetic particles can be come from limits on the strength of magnetic fields of real robust magnets that have to work for years. The probabilities of producing previously unknown particles or of finding previously unknown interactions necessarily decrease at higher energies, so higher-energy facilities must have higher intensity too. That's not obvious, but it's a consequence of quantum theory. Many innovative techniques to accelerate particles to higher energies provide only low intensities, so they are not useful for particle physics. We are not yet at the technological and financial limits of what we can do to probe nature more deeply using colliders, but we are nearing those limits.

One of the exciting aspects of the view of the world suggested by supersymmetry is that we may be able to get much of the data needed to formulate and test the final theory from the facilities already operational and those already being designed. The masses and production rates of the superpartners will tell us a great deal about the form the final theory takes—as will the nature of the dark matter of the universe, and possible tiny effects in the way neutrons and electrons behave in electric fields, and whether the proton is a stable particle or eventually decays, and more. We won't know until we succeed whether we really can have all the data we need to formulate and test the final theory. But for the first time in history, that we might do so in the foreseeable future is now a defendable argument.

Don't misunderstand. I am not claiming we know how to achieve such goals. Clever people can think of many things that could go wrong and of obstacles that could be too formidable. But such obstacles are not inevitable, and may not be there.

What if no signals are found for superpartners or other physics beyond the Standard Model, and the observed Higgs boson does not deviate from having properties like a Standard Model one? That is sometimes mentioned as a nightmare scenario. In a compactified M/string theory world that is actually possible, and it would be evidence for such a theory. We'll see later (specifically in Chapter 8 and the Appendix) that the compactified M/string theories do predict that the partners of quarks and leptons are probably heavier than about 30 TeV (thus too heavy to observe directly at a collider), and that the Higgs boson has properties close to those of the Standard Model. The gluinos and a light dark matter candidate wino-like superpartner, plus two more superpartners, are expected to be rather light and easily observable at LHC. The arguments for the heavy squarks and sleptons, and for the Standard Model–like Higgs boson, are quite generic and strong. The arguments for the gluino and wino being light are more subtle and complicated, and perhaps the latter will be heavier, too, when the theory is understood better. Then we would observe the nightmare scenario, but it would be implied by compactified M/string theory! It would be very interesting to see how many people could accept the indirect reasoning and see the "nightmare" as strong evidence for a supersymmetric compactified M/string theory world, with a gravitino* mass of order 30 TeV.

---

* See Glossary and Chapter 8.

6
· · · · · · · · · ·

What Is the Universe Made Of?

By the early 1980s, people recognized a surprising implication of a supersymmetric world with a stable lightest superpartner. It turned out that in a supersymmetric world after the big bang, the universe would contain many of the lightest superpartners—so many that their gravitational attraction for the other particles would have significantly slowed the expansion rate of the universe. The big bang was the creation of huge numbers of all of the basic particles, quarks and leptons and Ws and Zs and photons and gluons, and also superpartners. The particles moved throughout the expanding universe (which was then small, perhaps the size of a soccer ball), colliding with each other. The heavy superpartners were decaying, into the lighter ones and Standard Model particles. After a while, only stable particles were left, the up and down quarks, gluons, photons, electrons, neutrinos, and the lightest superpartners. We speak of all these particles as "relics" since they are all left over from the big bang itself, or from decays and annihilations of other particles into them, soon after the big bang. Essentially all the up and down quarks and electrons in us were made in the big bang, and have been around since, many in each of us.

Eventually, as the universe cooled, the quarks and gluons bound into neutrons and protons, and the neutrons and protons bound into the lightest nuclei (deuterium and lithium and the very stable helium). Joining with electrons, the nuclei formed atoms. One of the great successes of our understanding of the first few minutes of the universe is the prediction and subsequent confirmation that about 24 percent of the atoms in the universe are helium. Eventually clumps of hydrogen and helium condensed from mutual gravitational attraction and formed stars. Some of the atoms remain as "dust" spread throughout the universe. From observations of starlight and x-rays, we can deduce that a little over a fifth of the matter in the universe is in stars and planets and dust (about a twentieth of the total mass-energy density of the universe).

How do we learn how much total matter there is in the universe? Basically, astronomers observe the motion of stars in galaxies, galaxies in clusters of galaxies, and so on. All of them move in paths described by Newton's laws of motion and gravitation. The strength of the gravitational force on a star is proportional to the mass of whatever is attracting the star, and decreases if the star is further away. In the solar system, for example, this means that planets orbit the sun in longer "years" if they are further from the sun. One would expect a similar result for stars moving in galaxies, but it turns out that their speeds are much faster than expected. Analysis shows that rather than most of the mass of a galaxy being at the center as starlight would suggest, the mass is spread throughout the galaxy more or less uniformly.

Another approach to estimating the total matter content is to compare the recent expansion rate of the universe with the rate much earlier. If there is lots of mass, the recent rate will be slower than if there is only a little mass because of the gravitational attraction of the additional mass. One tool used to make these kinds of measurements is the "Doppler shift" of the light emitted by atoms. Every hydrogen atom in the universe emits light composed of a unique set of colors; the color of light is determined by its wavelength or, equivalently, its oscillation frequency. The way sound is shifted to a higher pitch (frequency) if the source of sound (e.g., an ambulance siren) approaches us is a familiar experience; the shift is larger when the speed of the approaching siren is larger. If the siren is receding away from us, the shift is the other way, to lower frequency. For light, the same effect occurs. An approaching source appears to be higher frequency, or bluer, than if it were stationary, and a receding source appears to be redder—it is red-shifted. The original discovery of the expanding universe came, in the 1920s, from observing that the light from distant stars is red-shifted. Thus, the amount the light is red-shifted at a given distance gives us information about how fast the universe is expanding.

A third approach to learning how much energy and matter there is in the universe is to study the energies of the photons left over from early in the history of the universe; their energies are affected by the other sources of gravity. Combining various kinds of information like these allows astronomers and cosmologists to deduce the total amount of matter in the universe.

WHAT PARTICLES ARE THERE IN THE UNIVERSE?

There are lots of photons in the universe, about 400 in every cubic centimeter. About 1 percent of the static on the screen of your TV when no channel is being received is due to those photons. The photons can have a range of energies. Another of the great predictions of the big bang picture that gave us confidence that we understand the big bang was the calculation and then observation in the 1960s of the number of photons of each energy.

Although the relic neutrinos have not yet been directly observed, and are very difficult to detect, the number of them can be calculated by the same methods that produce the right numbers of photons and of helium nuclei, so we can have confidence in the result. There are nearly as many neutrinos as photons. To deduce their effect on the expansion of the universe, we need to know not only how many there are but also their mass. Observations so far confirm that neutrinos have mass, but these observations give only a lower limit rather than a measurement of the mass. What is actually measured are the differences between the masses of two pairs of neutrinos, but the masses themselves have not yet been measured and are very difficult to measure. We expect neutrinos to have mass, since theories that extend the Standard Model normally predict that they do.

An interesting astronomical argument implies that at most about 20 percent of the matter in the universe could be neutrinos. That is because neutrinos are so light, and interact so weakly, that if all the matter in the universe were neutrinos they would just keep moving and never clump into galaxies as is observed. The fraction of the mass of galaxies that is neutrinos can be indirectly measured, and will give an upper limit on the amount of dark matter that is in the form of neutrinos. Altogether, current evidence suggests that neutrinos make up at least 1/4 percent and at most about 2 percent of the matter in the universe. Experiments now taking data or under construction could settle this issue over the next decade.

Combining all these forms of matter including the stars and dust, we see that at most about a fifth of the matter in the universe is accounted for. Whatever the rest is, it is not made of the particles that we are made of or even know about! In the early 1980s this accounting had not yet been done, and it was not recognized that some then unknown form of matter was needed to explain the total amount of matter. Once it was realized that the lightest superpartner would provide a form of matter that would affect the expansion rate of the universe, estimates were made of how much supersymmetric matter there might be, and what it would do to the expansion rate. This was already a significant test of the possible validity of supersymmetry, since it could have happened that for any reasonable lightest superpartner mass the universe would be "overclosed"; in other words, the lightest superpartner would exert so much gravitational attraction that the universe would stop expanding and collapse back long before it got to be the present age. Since you are reading this, that did not happen. In fact, it turned out that the lightest superpartners would give about the right amount of matter to account for the total amount, though until the mass and the interaction properties of the lightest superpartner are measured we won't know if this is true.

Because it interacts too weakly, superpartner matter would not join in the reactions that occur in stars to make starlight. The superpartners do have ordinary weak interactions with each other and with Standard Model particles, and they do cluster with other matter loosely into galaxies because all matter feels the gravitational force. If they were the only particles that existed, the universe would be dark, just as it would be with neutrinos, so these forms of matter are called "dark matter." The lightest superpartner is expected to have a mass about that of the W or top quark masses, very heavy compared to that of protons and light nuclei, so essentially all of its energy will be in mass rather than in motion. Higher temperatures correspond to faster motion of particles, so we speak of the lightest superpartners—which are approximately at rest—as "cold dark matter." Neutrinos will be much lighter than the lightest superpartners, so much more of their energy will be in motion; thus they are called "hot dark matter."

Some people assume or think that proponents of supersymmetry introduced the lightest superpartner, or even the idea of supersymmetry, to account for the dark matter. So it should be emphasized that the idea of superpartners and a stable lightest superpartner emerged without awareness of a connection to dark matter. The prediction that the lightest superpartners would form cold dark matter was made without knowledge of the need for a new kind of cold dark matter. When the prediction was made, it was known that there was more matter in the universe than could be accounted for by stars and dust, so something like dark matter was needed. But it was thought that the needed matter could be ordinary nuclei (consisting of protons and neutrons) in big planets like Jupiter, or white dwarf stars that no longer were visible, or that some of it could be neutrinos. Today we know that none of these can account for the needed dark matter; that is, if we add up all of the forms of matter predicted by the Standard Model, we cannot explain the observed amount of matter in the universe. The lightest superpartner could account for it. Even if supersymmetry provided only the right stuff to account for the majority of matter in the universe, it would be very important. It might do that, and account for the Higgs mechanism, and much more as well (as described in Chapter 4 and elaborated further in the following chapters), so naturally many physicists are very excited about it. At the same time we should keep in mind that this is frontier research, so new data and new ideas can change our understanding. For example, there are other possible forms of matter (not described in this book) that people have speculated could perhaps also account for some dark matter.

The lightest superpartner could be the partner of various Standard Model particles—for example, the photon ("photino"), the Z or W bosons that mediate neutral weak interactions ("zino" or "wino"), and others. In quantum theory the lightest superpartner could even be a mixture of several. Each of the mixtures interacts with quarks differently, and would annihilate with other LSPs differently, so if they are observed at the LHC or seen in direct or indirect detection experiments (which are described below), it will be possible to identify the type of superpartner that the LSP is. Underlying theories normally predict that the LSP is one or another type, so once that is identified it will help point toward the theory that extends the Standard Model.

IS THE LIGHTEST SUPERPARTNER
THE COLD DARK MATTER OF THE UNIVERSE?

How can we find out whether the lightest superpartner is indeed the cold dark matter of the universe? To be certain we will need to do two things: observe the lightest superpartner in laboratory experiments, and then measure its properties well enough to calculate how many lightest superpartners are left from the big bang, allowing for depletion by those that annihilated as the universe evolved. In Chapter 5 we surveyed ways that superpartners could be detected at colliders. Superpartner events have two lightest superpartners escaping the detector. Detection of the lightest superpartner at colliders involves indirect reasoning rather than direct observation, but it is only one more step in the chain of reasoning going from the photons bouncing off objects into our eyes to analysis of digitized information from detector modules. If the lightest superpartner is the cold dark matter, superpartners will be observed at colliders, and the lightest superpartner will be detected indirectly by this method. If such events occur we can proceed to learn about the LSP properties.

There is an interesting subtlety. To be the cold dark matter the lightest superpartner has to live at least as long as the lifetime of the universe (about 13.7 billion years), which it does if it is stable. But it is also conceivable that a particle could live long enough to escape from a detector (which takes only somewhat over a billionth of a second) but decay soon after (say, in a second or a year) and thus not be the cold dark matter. However, the typical lifetime for a superpartner is much, much shorter than a billionth of a second. Thus, if a superpartner escapes from the detector rather than decaying, it is living much, much longer than it would naturally live. That probably means that it is indeed stable. But for such an important question as what makes up most of the universe we want better evidence, so we need to find ways to show experimentally that the LSPs really live as long as the universe and are spread all over the universe today.

There are several methods being pursued to explicitly detect the lightest superpartner in a context that confirms its status as cold dark matter. The first is the most direct (and is often called "direct detection"). If the lightest superpartner is the cold dark matter, there are lots of lightest superpartners just sitting around in the universe. Our sun is moving around in the galaxy, and the earth is moving around the sun. Consequently, each of us, and a detector we build in a lab, are moving through a cloud of lightest superpartners that are more or less at rest in the universe. There might be one or a few in every region the size of a soccer ball. As we move through them, occasionally (rarely) a nucleus in us or our detector will interact with a lightest superpartner and bounce off it (recoil). So it is necessary to make detectors that are very sensitive to such a collision, which is very hard to do. Several approaches are being tried. One is to make a detector so that it is normally a superconductor, which conducts electricity with no resistance or loss of current. Superconductors are extremely sensitive to temperature, and the heat energy they gain from even one recoiling nucleus can make them lose their superconductivity, thus having a very large effect that is detectable. Making detectors that are sensitive to such small amounts of energy is very challenging, but it has been done after nearly three decades of development. Today there are about twenty detectors operating around the world that hope to detect signals of collisions with the dark matter particles, each using techniques that might give them an advantage depending on the properties of the dark matter. Once dark matter particles are found, data from different detectors can be combined to help unravel the properties of the dark matter. Such detectors are usually located underground to reduce the number of cosmic rays that could collide with nuclei in the detector and mimic dark matter collisions.

A problem for such experiments is that if a nucleus in the detector decays radioactively, the decay products also deposit energy in the detector, in an amount similar to the collision with the cold dark matter, so it can mimic a cold dark matter collision. If an experimental group reports a tentative signal, there are several possible ways to check that it is indeed from cold dark matter. For example, during half the year the earth and the sun move in the same direction so their speeds add, giving a larger velocity and therefore more frequent interactions. During the other half they move oppositely, so the speeds subtract, giving less frequent interactions; thus the event rate should vary in a predictable way over the course of a year.

Several direct detection experiments have reported possible signals. Because the techniques are new and there are possible ways to mimic signals, it is not yet clear which reported signals are valid and how to interpret them. Considerably more data from the direct detection experiments will come in the next few years.

A second method takes the detectors into outer space. Suppose there are indeed cold dark matter lightest superpartners all through the galaxy. Anywhere in the galaxy a pair might annihilate, giving ordinary quarks and antiquarks, electrons and positrons, and photons and neutrinos in the final state. The normal numbers of electrons and neutrinos and antiquarks and photons present in space are so large that an excess of those from lightest superpartner annihilation could not be detected. But positrons do not encounter an electron so quickly, so they might hang around a while, and an excess of positrons should be detectable. Antiquarks occasionally combine into an antiproton, which on the earth would annihilate quickly but also might hang around in space for a while and be detectable. There may be too many photons to see an excess in general, but an excess would be detectable for photons of a particular energy. Any of these signals are called "indirect detection."

So scientists have launched satellite detectors that can see an excess of antiprotons or positrons or photons of a particular energy. Evidence for a nonstandard positron excess has been reported by a mainly Italian group whose detector is called PAMELA. Another satellite called FERMI detects photons and, recently, an excess of photons having one particular energy. They also have detected photons with a certain range of energies so comparisons can be made with predictions. An even more ambitious approach is being pursued by a group that has put a large detector called AMS on the US space station. It should be able to detect excesses of antiprotons, positrons, and photons and to measure their energies. AMS has been taking data since July 2011 and should report results soon. At present, it is not settled whether all the reported effects are real and how to interpret them, but exciting new results about dark matter may be established before long.

There is another possible form of dark matter that is harder to detect. It is called "axions." These are unfamiliar particles with unfamiliar properties, with behavior that is hard to explain, but they could be important. They were first invented to provide a solution for why a possible interaction of quarks in the Standard Model was apparently not present. Later it was realized that they arise naturally in M/string theories. They can be very light, and they interact very little (far less than the Standard Model weak interactions), but it could happen that so many of them exist that they make up much of the dark matter.

It is possible to detect axions in a fascinating way using two of their properties. First, they can be emitted in large numbers by the sun. Second, an axion can make a transition into two photons, or equivalently a photon can absorb an axion and turn into a photon of a different energy. Imagine putting up a wall where the sun shines on it. Normally there would be no photons on the side of the wall away from the sun. Now imagine putting a large magnetic field on the side of the wall away from the sun. The magnetic field has a large number of low-energy photons. The wall is transparent to axions. One photon in the magnetic field could absorb an axion and turn into another energetic photon, so suddenly energetic photons would appear on the side of the wall away from the sun when the magnetic field was turned on, but not if the magnetic field was off. The energy of the photons would tell us about the properties of the axions. More realistic variations of this experiment are under way.

It's important to understand that none of the explicit detection methods will allow us to conclude that the particle detected is actually all of the cold dark matter, the full relic density. Both direct and indirect detection are from interactions of dark matter particles that have typically been around for billions of years, so either confirms that the detected particles make up at least some of the stable dark matter. If the dark matter candidates were observed only at colliders, we would not know that they had lifetimes at least as long as the age of the universe, since they would only need to escape the detector to appear to be stable. To calculate the actual amount of superpartner cold dark matter in the universe, we need to know the number of cold dark matter particles in the universe as well as their mass. Once there is a dark matter observation, the interpretation of the experimental results depends on those quantities, but also depends on other properties, such as the probability that the lightest superpartner will bounce off nuclei, or that it will annihilate, or occur in the decay of a superpartner produced at the LHC. Without some other way to measure or calculate these other properties, we cannot obtain all the needed information.

Of course if an effect is observed we will be happy, and perhaps convinced psychologically, but we will have to do better to really claim we know what constitutes most of the matter of the universe. It is surprising but true that even when we have detected a dark matter candidate we *cannot* measure how much of the total dark matter it comprises. Using as much information as possible we can estimate the answer with calculations, but there will always be some uncertainty, at least 10–15 percent even with the best data and calculations. Once we have a good underlying theory that relates ordinary matter and LSPs and any other forms of dark matter we will be able to sharpen the calculations, but still not fully determine the amount of dark matter precisely.

7

Why Is Higgs Physics So Exciting and Important?

EARLIER I BRIEFLY DESCRIBED the important and somewhat mysterious "Higgs physics" (known by the name of one of its inventors, Peter Higgs), and how physicists discovered experimentally one of the Higgs bosons that must be found if the Higgs physics ideas are correct. To confirm that what was discovered is indeed the Higgs boson, more data and analysis are required. To be "the Higgs boson" means it is the quantum of the Higgs field that quarks and leptons and W and Z bosons interact with to acquire mass in a way consistent with the theory of the full Standard Model. At the time of this writing, in November 2012, some people will still want to call the discovered particle "Higgs-like." Since the data are tentatively behaving as they should for a Higgs boson, we will for simplicity call it the Higgs boson. The discovery was scientifically and emotionally powerful, of interest not only to physicists but to many in the general public. It was a heroic theoretical and technological and cultural achievement, not only by individual scientists but by our society. Scientists figured out a major subtle and hidden aspect of how the physics universe works, designed the experimental hardware and analysis, and developed the technology needed to test the ideas. Society valued the knowledge and increased understanding enough to support constructing the multibillion-dollar giant experiments needed to test the ideas.

The Standard Model in its original formulation is a mathematically consistent theory only if all of the particles have no mass. If one just adds in masses arbitrarily the theory becomes inconsistent, and does so in a way that destroys the ability to do calculations and predict or interpret the results of observations. This difficult obstacle was overcome in a brilliant manner by several people in the 1960s who figured out how to add to the theory a hypothetical Higgs field that automatically interacted with all the particles that actually had mass, in just such a way as to allow the theory to stay consistent. The resulting theory could describe the masses of the particles, and allowed calculations of observables; the resulting predictions explained much that had been known but not understood up to then, and correctly predicted many experimental outcomes. A significant aspect of that achievement is that the W and Z bosons, and *also* the quarks and leptons, must get their mass from the Higgs mechanism. But for technical reasons the bosons and the fermions cannot get mass the same way. The Standard Model including the Higgs physics handles both bosons and fermions in simple and natural ways, without additional assumptions, which is a major success.

Technically the Higgs physics works perfectly, and there is little doubt that the Higgs physics provides a correct description of nature. But from the conceptual point of view this aspect of the Standard Model is not yet understood. While the Higgs physics of the Standard Model will probably not change, how we interpret it—and its implications for extending the Standard Model and strengthening its foundations—are still not closed issues. One of the exciting achievements of the supersymmetric Standard Model is that it provides an explanation of how the Higgs physics works, and the supersymmetric Standard Model embedded in broader theories (such as compactified string theories, discussed in the next chapter) can fully explain both the existence of Higgs physics and how it works. Some testable predictions that we will examine later follow from the supersymmetric explanation.

There are several reasons why the Higgs physics plays a special role. As mentioned in Chapter 2, the Higgs boson is a previously unknown *kind* of particle. Electrons and quarks are the matter particles (they all are fermions). They are all really one kind of particle, differing only in that they carry different amounts of the various charges—electric charge and strong charge and weak charge. Similarly, the bosons that mediate the Standard Model forces also come in several varieties—photons, gluons, and W and Z bosons. They are all like the photon, differing only in how they interact with the fermions and with each other. The fields of which these particles are quanta are all very much like the more familiar electric and magnetic fields whose quanta are the photons. The Higgs bosons, on the other hand, are the quanta of a Higgs field whose origin is not yet understood, a field that seems to be quite different from the other fields and from our experience with fields.

Another perhaps profound role for the Higgs boson arises because it allows electrons that make up atoms to have mass. If electrons were massless, atoms would have effectively infinite size, and small objects like people and planets would not exist.

The universe, like any system, is expected to sit in its lowest energy state, called the ground state, or vacuum. Fields have energy, so one would expect that in the lowest state the fields are zero. That's true for the electric and magnetic fields, but not for the Higgs field: in the vacuum, the Higgs field has a nonzero value. (In case you want to read more technical literature, the jargon is "vacuum expectation value," commonly abbreviated as "vev.") This is not just the usual statement about the vacuum—namely, that quantum fluctuations occur all the time, with quanta or pairs of quanta fluctuating out of nothing and back to nothing. That occurs, but for the Higgs field it is the "nothing" that is changed. The vacuum state of the universe is filled with a nonzero Higgs field. The Higgs field carries charge, too—not electric charge but the so-called weak charges of the Standard Model. The

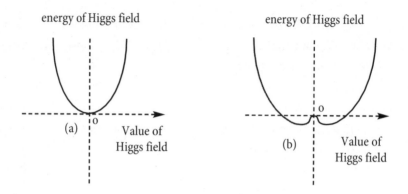

Figure 7.1

Higgs field is not zero and it is not neutral. Specifically, particles with electric charge give rise to electromagnetic fields, but the Higgs field does not have a source.

Pictorially one can think of the Higgs mechanism as in Figure 7.1. The curves represent the energy of the Higgs field. In 7.1(a) the field will settle at the bottom, with the lowest energy. The horizontal axis represents the value of the Higgs field in the lowest energy state, so in 7.1(a) that is at the origin and the Higgs field has zero energy when its value is zero, similarly to other kinds of fields. In 7.1(b) the Higgs mechanism makes the shape shown, like the bottom of a wine bottle. If the field were at the origin it would be at the top of the bump and roll off, ending up in the trough, with a nonzero value of the Higgs field in the ground state, the vacuum. In a supersymmetric world the shape is derived to be as in 7.1(b), and why that is nature's solution can be explained.

THE HIGGS FIELD, MECHANISM, AND BOSON

If you want to follow the discussion in detail, it's important to distinguish three aspects of Higgs physics. First, there has to be a Higgs field at all, and eventually its origin has to be understood. Second, all fields have quanta, so quanta of the Higgs field will exist. They are the Higgs bosons. Third, that field has to be nonzero in the ground state of the universe. That's called the Higgs mechanism. In the Standard Model the Higgs mechanism is simply assumed; in fact, it cannot be derived or explained in the Standard Model. One of the reasons that many people got excited about having the world be supersymmetric, in the early 1980s, is that in a supersymmetric world the Higgs mechanism can be derived, so the nonzero vacuum value of the Higgs field is explained. The interactions of the Higgs boson allow us to learn if there is indeed a Higgs mechanism. It's actually even better, because the supersymmetric derivation of the Higgs mechanism is a calculable so-called perturbative one, for which we already know how to do the calculations, rather than requiring a complicated and difficult mathematical analysis.

Two unambiguous main tests that the observed boson is indeed a Higgs boson are, first, that the particle should have spin zero and, second, that there should be interactions where the Higgs boson makes a transition into a pair of Z bosons and also a pair of W bosons, with the strengths of these transitions being of a certain predicted size and proportional to their masses squared since they get mass by interaction with the Higgs boson. That has now been observed in the LHC data!

Remarkably, the Higgs boson found at the LHC looks like a pure Standard Model Higgs boson in its decays (even though it cannot be because the pure Standard Model one gets the huge quantum corrections that I mentioned earlier). This means that its observable decay rates seem to be about the same as they would be if it were a Standard Model one, though they need not have been the same. More data and analysis are needed to see that this is indeed true. People seldom mention that this property is surprising. As explained in more detail in the next chapter and in the Appendix, this feature and also the value of the Higgs boson mass were both predicted (before the data) by compactified M/string theories! The actual numerical value of the mass provides significant information about the underlying theory and favors an M/string theory interpretation. Data are very powerful—this did not have to be the case.

The fraction of Higgs boson decays to any particles that get mass from their interaction with the Higgs field should be proportional to their mass (actually, the mass squared). That's expected for all the channels except the decay to γγ, since the massless photons do not get mass from interacting with the Higgs field. The decay to γγ is via some intermediate particles, and is smaller than the rest of the modes, but still predictable. Because it occurs via intermediate particles, it is sensitive to additional contributions. The experimenters designed the parts of the detectors that capture and measure the energies of the photons to optimize the Higgs boson detection via the γγ decay.

It is important to measure as many properties of the discovered Higgs boson as possible, as well as possible to be sure it behaves like a Higgs boson should. One such property is its spin—Higgs bosons must have spin zero. The evidence is already good that this is indeed so. Spin is quantized, and here only spin zero, one, and two are possible. Spins larger than two would predict different production rates, but the observed rate is close to the expected one. It can be derived from the rules of quantum theory that if a particle decays into two photons, as the observed one does, then it cannot have spin one. The Higgs signal is also observed with about the predicted rate at the Fermi National Accelerator Laboratory Tevatron collider near Chicago, via its decay to b quarks. ATLAS and CMS are reporting signals of decays into b quarks or tau leptons. Either of these decay modes implies that the spin is zero or one, but spin one has been excluded by the decay to two photons. So data already imply that the spin is zero. When more data are available, some additional checks can be done to confirm that the spin of the observed Higgs boson is indeed zero.

Each of the Higgs boson decays plays a significant role in teaching us about its properties. Theoretically the decays to ZZ* and $W^+ W^-$ are the most important ones to observe, since they teach us that the electroweak symmetry is broken and the Higgs mechanism is operating. That's complicated to explain technically, but basically the electroweak symmetry of the Standard Model would forbid the Higgs boson to have those decay channels, and their presence shows that the symmetry is

broken in the right way to allow the interaction that gives mass. The *
on Z or W indicates that the Higgs boson is too light to decay into two
real Zs or two real Ws, so one of them has to be a virtual particle: such
a decay is understood theoretically, and how to deal with it experimentally is known.

The relative sizes of the different decay modes tests that the Higgs
interaction strength is determined by the mass of the various particles.
Soon (presumably by the time this book is published) there should be
enough data to confirm that for the decays mentioned above. A good
check later will be that the decay to $\mu^+ \mu^-$, a very clean one experimentally, should be much smaller (about 300 times smaller) than the $\tau^+ \tau^-$
mode since the squares of their masses are in that ratio. A probable signal in the τ channel has been reported, so it will first be a good check
that no signal should be reported in the μ channel.

The role of the Higgs physics in the Standard Model is a necessary
one. Physicists can be divided into three categories according to their
attitude about the form the Higgs physics will take. Some, the funda-
mentalists, think it is a fundamental particle, the Higgs boson, and that
it has been found. That is what supersymmetry predicts. Others believe
there is no fundamental particle at all, but that some as yet unknown
form of the interactions among particles at higher energies will somehow manage to look just like the Higgs-like particle discovered—they
are the atheists. Some atheists have invented very sophisticated models
with unknown quarks and leptons interacting with unknown forces in
just such a way that some of them bind into bosons that are just like
Higgs bosons, all to avoid having Higgs bosons that are as fundamental
as photons and electrons. In 1993, one of us edited a book, *Perspectives
in Higgs Physics*. Representatives from all points of view were invited to
contribute chapters. One Higgs atheist, Howard Georgi, was offered a
title that he agreed to: "Why I would be very sad if a Higgs boson were
discovered." The third group are agnostics, unsure. The arguments
among the proponents of the groups have been fruitful ones because
they have led to thinking of and working out better experimental and
theoretical ways to learn how nature actually solved the Higgs physics

problem. The experimentally observed Higgs boson not only completes the Standard Model, its properties will help us know how to extend the Standard Model. Of course, if superpartners are discovered at LHC we will know that the route to a deeper understanding is via supersymmetry.

NOT THE STANDARD MODEL HIGGS BOSON

Over three decades ago people noticed that the pure Standard Model Higgs boson could not be the entire answer, because one could prove that in a quantum field theory the numerical value of its mass would get large quantum corrections that made its mass a huge value. The masses of quarks and leptons and W, Z are proportional to the Higgs mass, and would also be huge, far larger than the observed ones. A pure "Standard Model Higgs boson" cannot exist in a consistent mathematical theory. Although people continue to speak of the Standard Model Higgs boson, and such a construct is useful for figuring out ways it could decay and ways it could be detected, the final answer for the Higgs physics must be different. This is called "the hierarchy problem."

It was recognized in the early 1980s that the Higgs bosons of the supersymmetric theory did not have the problem of getting huge masses in quantum theory, and could give a consistent solution. Although the supersymmetric theory existed before this problem was known, that the supersymmetric theory could solve the hierarchy problem soon came to be seen as another major motivation to study the supersymmetric theory and to think it might be correct. There are actually five Higgs bosons predicted by the supersymmetric theory. They can and typically would have masses and decay rates different from the Standard Model one. All should be found in experiments eventually if colliders are built that have enough energy to produce them.

Making scientific discoveries at the frontiers requires expensive facilities, innovative technology (which is why it is always a good investment), and strong commitment and leadership. The history of the search for the Higgs boson provides a lesson in this area. As soon as the

Superconducting SuperCollider was cancelled by the US Congress in 1993, I and other theorists recognized (and pointed out) that the Higgs boson could be detected at the Tevatron collider at Fermilab if already-planned upgrades in its energy and intensity were carried out, if the detectors were state of the art, and if the mass of the Higgs boson was in the range allowed in supersymmetric theories (as it turned out to be). The reported Fermilab confirmation of the LHC discovery via the b quark Higgs boson decay demonstrates that the discovery was indeed possible at Fermilab. It was essentially a matter of achieving enough collisions, keeping the detectors functioning well enough, and having manpower to analyze the data. In fact, the Higgs search at Fermilab was given too little priority and government funding. The discovery of the Higgs boson could have been made at Fermilab several years ago. It's interesting to think about whether that matters.

The reader may note that I have not used the common analogy of something moving through molasses or a crowd being slowed down to explain how particles get mass from the Higgs field. That's because it's rather misleading and technically wrong. Interactions with fields give accelerations (think of an electron in an electric field), not velocity, and it doesn't explain why electrons get mass but not photons, for example. Unfortunately there aren't any good analogies known so far, and the molasses or crowd one is probably better than nothing.

In any case, the discovery has now been achieved so we will turn to its impact on our search for the underlying laws of nature, particularly for achieving an understanding of our universe (our "M/string vacuum") and perhaps going beyond that. The next chapter introduces M/string theory and its role, and in the Appendix the compactified M/string theory prediction of the Higgs boson mass and properties is described in somewhat more detail for interested readers.

8

. . . .

M/String Theory!

WHATEVER YOU MIGHT HAVE HEARD to the contrary, the development of M/string theory is wonderful progress in physics. I'll explain why in this chapter. M/string theory provides a framework in which most or all of what we want to understand and know about our physical universe, from the smallest scales to the largest, can be addressed. Even better, the questions can be addressed in a unified way, with answers to any one question related to answers to others. M/string theory also provides a consistent quantum theory of gravity, which is important.

There are a lot of magazine articles and books and blogs that argue M/string theory is not testable and even not science. That's wrong, as I'll explain. One common mistake is to assume that you have to be somewhere to test ideas about there. M/string theory is naturally formulated at the Planck scale (sizes of order 10^{-33} centimeters, energies of order 10^{14} times the LHC energy). In this book we don't consider alternatives to this simple, natural picture. Obviously doing experiments at the Planck scale is not possible technically or financially. But it is normal in physics and science in general to test ideas based on their predictions. Nobody was around during the big bang, but knowledgeable people are confident it happened, because several predictions based on the big bang physics have been verified. The three most dramatic ones are the expanding universe, the correct prediction of the abundances of the light nuclei (called nucleosynthesis) in the universe, and the cosmic microwave background radiation. Testable M/string theory predictions have and are being made—although, surprisingly, making such predictions is not a major priority among M/string theorists, many of whom prefer to work on the ten- or eleven-dimensional theory for its own sake. The predictions require LHC data or satellite data and only recently could be tested. The mass and Standard Model–like behavior of the Higgs boson were correctly predicted before the data. (I describe how the prediction works in the Appendix.)

We live in three space dimensions. String theory has to be formulated in nine space dimensions or it is not a consistent mathematical theory. There doesn't seem to be a simple way to explain "Why nine?" What happens is that if theories to describe nature and to include gravity in the description are formulated in d space dimensions, they lead to results that include terms that are infinite, but the terms that are infinite are multiplied by a factor $(d–9)$, and drop out only for $d = 9$. Although we seem to live in three space dimensions—and after Einstein's special relativity, which mixes time in, we sometimes refer to four dimensions—physicists have for many years looked at theories in different numbers of space-time dimensions. Here we imagine d dimensions, without immediately specifying d. To make a testable prediction, we need to formulate the theory with the full "extra dimensions" if d is larger than three space dimensions, and then project the theory onto our three dimensions. That process is called "compactification." For string theory we then have $d = 10$ space-time dimensions.

String theory is very mathematical and has a lot of technical details and terminology, most of which the reader does not need to know to understand what is important. Some of the physics of string theory leads to procedures or quantities that do not have simple analogies in our experience—not even in our experience with the Standard Model.

As always, the words used to describe new procedures or particles are assigned when they arise and there are no guiding rules. The reader does not need to know most of them. However, three that would be good to add to one's vocabulary are "compactification," "moduli," and "generic." (The first of these was explained above, and the latter two are discussed below; see the Glossary, too.)

Another feature that is not important for our present purposes but should be mentioned is that there are several types of M/string theories, with names like Type II, Heterotic, and so on. They arise when a more general theory with eleven space-time dimensions, called M-theory, is projected onto ten-dimensional theories in various ways. M-theory is not literally a string theory. But the crucial thing is that, for our purposes, in connecting to the real world it is the compactified theories that matter. M-theory can be compactified to four dimensions too, and the compactified M-theory behaves like a typical compactified string theory. We refer to the compactified theories as "M/string" theories. We don't want to lose sight of the M-theory version by just calling them "string" theories, since there are already significant results from the compactified M-theory. It is not yet known if predictions for observable quantities such as the Higgs boson mass and properties, or the top quark mass (etc.), are different in different types of compactified M/string theories, although there seem to be cases where they may be. If they are different, then data will be telling us which compactification is the one that nature used for our M/string vacuum.

Today there are compactified M/string theories where each of the features of the Standard Model, and the most important beyond-the-Standard Model questions, can be derived and explained. There has been steady progress in working these out. Each compactified M/string theory has both explanations of existing results and predictions for results that can be tested at the LHC and in dark matter experiments. There is not yet a particular one where all the questions are answered. Once each question can be answered in some of the theories, it seems very likely that soon we will have testable examples where all the questions have been addressed or answered.

After compactification, in string theories we think of six small dimensions, curled up to be about Planck-scale size, and our four large dimensions. The curled-up six dimensions form a space that can be complicated, and can be described mathematically, called a manifold. M-theory has eleven dimensions, so four stay large and seven are small and curled up. The manifolds have an overall volume, and some of their dimensions might be larger than others. To describe the manifold mathematically, one has to provide information about such sizes and shapes. When the theory is compactified, information from the 10D or 11D structure is not lost; rather, such information determines relations among what the theory predicts—namely, the particle properties, masses, and interactions. The sizes and orientations of the curled-up regions are described by quantum fields, as is everything in a quantum theory. The jargon name for them is "moduli fields," and they have quanta called "moduli bosons." When talking and in papers, physicists usually just call both the fields and the quanta "moduli" and the context makes clear what is meant. There is no good analogy to help understand moduli better—they do not occur in our normal experience. But they do play an important role in M/string theory predictions, and I will use them a couple of times in explanations.

The extra dimensions are expected to be so small we do not perceive them directly, though their presence affects the form of the compactified four-dimensional theory. We perceive our familiar four dimensions as objects, including us, and events (the time dimension) move in them. The particles (or strings) can also move in the other dimensions. Similarly to how the usual dimensions define the 3D sizes of a button and how long it takes to sew it on a shirt, the others define the vibrations that nature uses as particles, which it then uses to make the atoms, which make the molecules, which make what's around us. There are expected to be several extra dimensions, and they can take on more shapes than the space we are used to that surrounds 4D objects and events.

The natural size for the extra dimensions is the Planck scale, described earlier. Some theorists have considered the possibility that some or all of the extra dimensions could be larger than that, which is logically possible. In this book we focus only on the approach that the extra dimensions are typically near the Planck-scale size.

Today we have to study compactifications one at a time and check that they are consistent with all the results of the Standard Model and all constraints from other particle physics data (such as quark masses, parity violation, etc.), and with all cosmological constraints (such as nucleosynthesis, the dark matter, etc.). Eventually perhaps there will be some way to have the ten- or eleven-dimensional theory point to the correct compactification. But before then it will be extraordinary progress if we can find an example where all the traditional questions of physics and cosmology have one unified answer, even if we only guess the compactifications to try. There has been considerable progress toward that goal in recent years. In nature any system will naturally fall into its state of lowest energy, whether it is a ball bouncing down hillsides to the bottom, or an atom, or a universe. In atomic physics that is called the ground state. In particle physics and cosmology it is called the vacuum, the state of lowest energy. In particular, we will call it "our M/string vacuum." Perhaps there are other, even many, M/string vacua. We will be able to figure that out, perhaps in parallel with finding our own M/string vacuum.

Most M/string theorists are more interested in M/string theory for its own sake than in the compactified theories and their connections to data, predictions, and explanations. There are an increasing number of people working in the increasingly active area called M/string phenomenology. The 11th International M/string Phenomenology conference was held at the University of Cambridge in England in July 2012, with about 125 participants. Also, the US National Science Foundation funded a three-year "M/string Vacuum Project" grant to a network of eight universities, with all funds going to help support PhD students in M/string phenomenology. The name is meant to emphasize that the focus in M/string phenomenology is on research about compactification and the ground-state solution of the M/string theories (the vacuum) rather than on the full 10D or 11D M/string theories.

WHAT IS M/STRING THEORY?

What is M/string theory anyhow? Historically, in physics, theories developed one piece at a time, with pieces getting pasted together and unified until they were rather complete effective theories. Eventually they got formal logical definitions. We sometimes call that process bottom-up. That's what happened with quantum theory. Sometimes people try to go the opposite way, trying to write the full theory all at once, top-down. M/string theory does not yet have a top-down definition. That's all right. I can give a definition of M/string theory that is the relevant one for understanding its importance and power.

What is any theory? The goal of physics is to write a consistent mathematical theory that describes the physical world, and to understand why that is the actual theory that describes nature. Hopefully it will have few or even no inputs (in the sense of the effective theories discussed in Chapter 3). It must be a quantum theory, and relativistic. The Standard Model is a consistent relativistic quantum field theory of the weak, strong, and electromagnetic forces that treats the particles as point-like objects. It works well. But we learned that a relativistic quantum field theory of gravity based on point-like particles leads to some meaningless predictions. M/string theory describes particles not as points but, rather, with the equations that would describe the motion and interactions of strings. That seems to work. It gives a theory of gravity with meaningful results, and it can describe all the particles and forces in a unified, mathematically consistent way in ten or eleven space-time dimensions. Probably any description in terms of objects more complicated than points would work and give the same compactified theories. In string theory an electron is still an electron, but it is described by the equations for a vibrating, interacting string rather than by the equations for a point.

We have seen that compactified M/string theories can be tested. More generally, what does it mean to test theories in physics (or science)? Consider $F = ma$, Newton's second law. Can it be tested? The answer actually is no. For some particular force F and mass m, the second law can be used to calculate the acceleration a (which can also be measured) and to test the validity. But only for one force at a time, not more generally. That's essentially the same way compactified M/string theories can be tested. So M/string theory is as testable as $F = ma$. A similar comparison can be made to the Schrödinger equation in quantum theory that is used to calculate the properties of quantum systems.

Let's elaborate on the goals of particle physics. What we want is first to learn what the laws of nature are, and if they can be unified into one or very few basic principles. We want to know why the laws of physics are what they are. We want to find the final theory (again, recall the discussion of effective theories in Chapter 3). We want to explain our world in particular, why it's there, and why it is the way it is. In order to do all of that, we not only have to understand why the particles and the forces are what they are, and the rules, but we have to learn what spacetime is, and why nature's rules for calculating the behavior of particles are quantum theory and relativistic invariance. We have to understand whether the final theory is unique, the only possible theory of its kind that could be mathematically consistent.

As was explained in Chapters 2 and 3, we have basically succeeded in understanding how our world works at all scales essentially down to scales of about 10^{-17} centimeters, and in a supersymmetric world possibly even near the Planck scale. We know the basic particles, the quarks and leptons, and we know the forces and how they operate, mediated by the exchange of the bosons W, Z, gluon, and graviton. We do not yet know why the world works this way. It's important to be aware of this distinction between "how understanding" and "why understanding." In physics the "how" came first, and now we want to have the "why." That order is inevitable in physics, given that we start with the world we see, but it can be different in other areas of science. For example, in biology the fundamental organizing principles (the "why") are two: evolution by natural selection, and the genetic code based on chemical molecules. All of the diversity and properties of life on Earth can be traced back to the first self-replicating molecules, evolving via differential survival of organisms in a changing environment. The changes in organisms can be traced to changes in their genetic structure. However, (for example) we do not yet understand in molecular detail how an egg plus a sperm develops into a person with eyes and fingers and a brain, though much has been learned about human development in recent years. Biology has the "why," the basic framework, but how the principles actually lead to the phenomena we observe is still being worked out.

Hidden or Broken or Partial Supersymmetry

In order for M/string theory to have a stable, long-lived vacuum, it is probably necessary for it to be supersymmetric. It must have full supersymmetry, not a broken supersymmetry. This is a very important distinction. The ten-dimensional M/string theory near the Planck scale is fully supersymmetric, and the four-dimensional supersymmetric Standard Model at the collider scale has the supersymmetry hidden or broken. When the supersymmetry is hidden it is not totally invisible: the superpartners still exist, but the particles and their superpartners can have different masses. Different ways of breaking the supersymmetry lead to different patterns of masses and interactions, so when the masses and interactions are measured we will have experimental information pointing to how the supersymmetry is broken.

In the supersymmetric theory the graviton, the quantum of the gravitational field, has a superpartner gravitino. Both are massless in the unbroken theory, but the gravitino becomes massive when the supersymmetry is broken. The gravitino mass then characterizes the size of supersymmetry breaking, and all the other superpartner masses are proportional to it. There are theoretical arguments that the gravitino mass is about 50 TeV (say, to a factor of two accuracy), but this is an active area of research and the question is not yet settled.

Why or how the supersymmetry is broken is not yet known, though there are known possible mechanisms. It was necessary to break the particle-interchange symmetry of the Standard Model before the Standard Model could correctly incorporate mass into its description. Figuring out how that worked was difficult, and resulted in introducing the Higgs physics. An analogous effort may be needed to account for the supersymmetry breaking.

THE ROLE OF DATA

M/string theorists work at the Planck scale, and their theories have unbroken supersymmetry, though they keep an eye open for how the supersymmetry might be broken. There are ideas and examples that may explain how the full supersymmetry at the Planck scale could become a broken or partial supersymmetry at larger distances. We don't yet have data on the superpartners, so we have no way of being sure whether one of the already-discovered methods to break supersymmetry or curl up extra dimensions is actually right. Supersymmetry theorists think about physics at the collider scale, working out how to recognize that superpartners have been produced and how to take the quantities experimenters actually measure, such as the probabilities for various processes to occur and the directions and energies of the particles that emerge into the detectors, and convert that information into the patterns that can be compared with possible implications of M/string theory.

As noted, the (expected) discovery of experimental evidence for (broken) supersymmetry and the data about the properties of superpartners will be superimportant not only because they teach us about the validity of various theories of our M/string theory vacuum but also because they open the window to view the physics at tiny distances like the Planck scale. Making the connection from the ten- or eleven-dimensional unbroken-supersymmetric M/string theory near the Planck scale to the four-dimensional compactified broken-supersymmetric Standard Model at the collider scale probably requires supersymmetry. Recognizing supersymmetry in the data and interpreting the form it takes will be important for getting to the theory of our M/string vacuum.

Basically theorists will "guess" compactifications that could give our M/string vacuum and, guided by the data, will choose promising ones.

Then they will calculate as well as possible predictions for observables that will soon be measured. If there is agreement, one goes on to the next observable in the same compactification. If there is disagreement, the compactification will be changed and we will try again. One might think there are so many compactifications that there will not be time to go through them all, but we already have enough experience to know that many observables and tests can be quickly checked for large classes of compactifications.

Sometimes people ask if M/string theory can ever be disproved. That's not actually how science usually works. Predictions depend on assumptions, and a wrong prediction usually tells us to change the assumptions. Eventually one runs out of assumptions and people lose interest in some approach. Theories are discarded, not disproved. Many people work on ten- or eleven-dimensional M/string theories and on their mathematical properties. To connect to the real world and to test M/string theories, we must test compactified theories. That can be done one at a time, but the situation is better. Some compactified M/string theory predictions apply to broad generic* statements about large classes of compactified M/string theories. For example, any compactified M/string theory will generically have moduli.* The moduli take on definite values in the vacuum, and if at least one them has a value determined by effects of broken supersymmetry (normally they all do) and if the theory is consistent with cosmology data, then it predicts that squarks will be heavy (more than 100 times the top quark)— too heavy to be detected at the LHC. Gluinos and neutralinos and charginos are expected to be observable at the LHC; the argument does not apply to them. If squarks are found at the LHC, a very large class of compactified M/string theories will be wrong. Note that this is a robust prediction from M/string theory for experiments, an example from a number of testable predictions. If no compactified M/string theories

---

* See Glossary.

emerge that describe our world well, then people will simply lose interest and stop trying. This is how physics has worked for four centuries—it is the normal path.

Anyone who was attracted to this book knows that particle physics and cosmology are entering a very exciting era because there will be data from the CERN LHC and from dark matter satellite and laboratory detection experiments and perhaps more. There is a second, less appreciated, reason: finally there is a theoretical framework to address the basic questions we want to ask, questions about the particles and forces, how they fit into a deeper and broader framework, and why they are what they are—namely, M/string theories. Without the framework, the experimental results would be puzzling unconnected facts. Today, with both exciting new data coming and the M/string theory framework, we can expect rapid progress.

9

···· ···

How Much Can We Understand?

SUPPOSE THAT SUPERPARTNERS are indeed detected and studied experimentally soon. Suppose that from their properties we are led to an M/string theory whose four-dimensional form explains the features we think are essential for understanding the world of particles and cosmology. How much further can we hope to go in understanding where the laws of nature and the universe came from—are there limits? Even if we can eventually understand the universe, is it premature to hope to do so soon?

As always with science, we can't know what we can understand until we get there, so it is only if we succeed that we'll be certain of the answers to these questions. But we can look at the arguments against the possibility of learning and testing the final theory and see how good they seem to be. The outcome is that the skepticism is not backed up by sound arguments.

First consider whether we can hope to discover the final theory. Remember, our current goal is to *understand our* world. Both italicized words are crucial. People have raised a number of issues meant to suggest that we are unlikely to achieve that understanding. One argument is that there is much that we cannot know about the world, and science has added to that list. For example, because of the finite speed of light and information, we cannot know what is happening right now on Mars, or halfway across the universe. We cannot know the position and momentum of a particle simultaneously to better-than-some accuracy set by the uncertainty principle. But our goal is not to *know everything* about the universe but, rather, to understand how the universe works and why it is the way it is. Learning that there are things we cannot know (as in the above examples) is actually part of that understanding, and in no way implies that we cannot understand how the universe works and why it is the way it is. We cannot know all possible chemical molecules, but we can fully understand the principles that govern the formation and behavior of all possible molecules. Understanding whether there are many possible universes can be studied in parallel with understanding our world—both are important.

Another argument is based on Gödel's incompleteness theorem. That is an astonishing and elegant mathematical result proved by Kurt Gödel in the 1930s. It basically states that in any mathematical theory that is interesting for our purposes, there are true results that cannot be mathematically proven to be true, and also that one can't prove the consistency of a mathematical system from within that system. For mathematicians that has profound implications. A number of people have worried that it also meant that physicists would be unable to show that the final theory applied to the world even if it did. But that is not how science works. There are two important differences. First, we do not need to prove all possible theorems, nor do we have to prove the consistency of the whole system. We already have our world, and we know it is described by consistent laws—otherwise it would fall apart. If the equations for the stability of atoms changed with time, or were inconsistent with those for the forces, atoms would not keep existing and forming the world. Only consistent equations have solutions. Indeed, it is often remarked that it is "amazing" that our world is comprehensible scientifically. But that is really not surprising. Our world must behave according to mathematical regularities if it is to exist for some time. Once that is the case, we can learn what the regularities are.

Second, scientific results are never proven to be true. "Proof" is for mathematical theorems. At a certain stage of research, the evidence for a given physics result becomes so strong it is accepted by those who understand it, and by others who trust them. Every result depends on certain parameters (such as distance or speed or others). If the evidence comes only from a limited range of those parameters, the result may or may not change when it is extrapolated outside that range; recall the discussion of Chapter 3. For example, the laws of gravity have not been tested for some distances smaller than a fraction of a millimeter. Experiments, motivated by some ideas from M/string theory and large extra dimensions, are under way to find out if the form of the gravitational force changes at smaller distances. On the other hand, our descriptions of all forces have now been tested for all speeds, from rest to the maximum possible speed (of light), and so no further modification will happen there. In addition, as we have seen, accepted scientific results form a coherent structure with many implications that strengthen our confidence in them. If one part is modified, the whole structure might change and contradict results that are related only when there is a theory. The results of science lead to our surest knowledge about our world because of the process used to obtain them, including improved experimentation and consensus of informed workers. The bottom line is that Gödel's theorem is simply not relevant to whether we can understand how our particular world works and why it is the way it is.

Yet another concern arises from a feeling of humility in the face of the awesome size and complexity and beauty of the universe, from the particles to the cosmos. How can mere humans expect or even hope to understand all of that and how it originated and why it is the way it is? Sometimes this is stated alongside the comment that it is like expecting a dog to understand quantum theory. Charles Darwin first used that analogy; he wrote, "We feel most deeply that the whole subject is too profound for the human intellect. A dog might as well speculate on the mind of Newton." It is easy to understand why Darwin would feel that way, living as he did before all that has been learned in physics over the past century and a half. We won't know if humans can figure out the final theory unless our efforts to understand either succeed or hit a dead end. The approach of science since it began has been simply to try to understand natural phenomena as well as possible, and to see how far we can go. Personally, I think that if a dog were able to get data about atoms and ask about how they work, then it could discover quantum theory, so I remain hopeful that we will indeed understand the universe.

TESTING STRING THEORY AND THE FINAL THEORY

Could we fail to understand the world because we cannot test ideas? This is a more subtle and interesting point. Some people have suggested that because we can never build a collider that can directly probe the Planck scale, we can never test ideas about physics there, or test the final theory. That is simply wrong, as we saw in the previous chapter. We do not have to be there when the dinosaurs become extinct to figure out how that happened—again, there are clues and relics that let us unravel the mystery. We do not have to travel near the speed of light to figure out and confirm that it is the maximum speed at which we can travel. There are always clues and relics that will help us test ideas. Finding those clues and relics requires dedicated effort by talented scientists, and financial long-term commitments from governments. It doesn't happen by accident; first ideas and hypotheses have to be formulated, and then tests emerge. People argued incorrectly in the past that it was not possible to demonstrate that atoms existed, or that neutrinos existed.

Similarly, M/string theory can address a number of questions that the Standard Model cannot address, such as the number of families of quarks and leptons, the values of the masses of the quarks and leptons, which phenomena should show "CP violation," whether protons decay, and more. If M/string theory incorporates quantum theory and gravity and the Standard Model forces into one description in a consistent way, and also explains why there are three families of quarks and leptons and calculates some of their masses from formulas that have no adjustable parameters, we will surely have the understanding of our world that we hope for. The problem is that the calculations needed to be sure M/string theory does all of that are very difficult, and they depend on how the small dimensions curl up and on which of many solutions nature actually selected. So even if M/string theory were indeed right, we would not know it until people are able to do the calculations and confirm that M/string theory does indeed explain the unsolved problems. Such work is under way.

The problem is partly psychological. For example, the development of a quantum theory of electromagnetism was hindered for some time because a number of calculations of observables gave infinite results when they shouldn't have. Then, in 1947, a measurement was reported for one of those observables, and soon after that, theorists figured out how to make the calculations finite and meaningful and got the right answer. What changed was that they took it more seriously when there were data, and they had a definite experimental answer to let them know if the result of their calculation was right. With M/string theory the question is whether it will be taken seriously enough so that talented people will focus on the calculations that test the ideas, even when that means investing years in work that might not pan out. Hopefully the situation will be better with the Higgs boson mass now measured and calculable, so that it can push theorists to understand better how to derive more predictions.

The situation will be similar for the final theory. If a candidate theory allows us to understand why the rules that describe nature are quantum theory and relativistic invariance, and to define space and time in a consistent way, we will not have much doubt that it is the answer. There may be additional tests there too, though until we get there we won't know what the tests are. Many of the tests of any idea emerge only after the idea is formulated and its implications are studied (as happened with the tests of the big bang or the prediction of radio waves from Maxwell's equations, for example). One can imagine some possible tests of the final theory. For example, there is a symmetry that must hold in any relativistically invariant quantum field theory, but which need not hold for the 10/11D final theory, and one can imagine ways to test whether or not it does. (Basically it states that if there is a physical process, then there must be another physical process giving identical results, obtained by taking the first one and changing all of its particles into their antiparticles, reflecting in a mirror, and running it backward in time.) Once we understand whether and why quantum theory has to be the way nature works, it may turn out that there are necessary modifications to its formulation that appear only at the Planck scale, and that can be searched for when we know the associated predictions (but this is unlikely). The key point is that since we can describe some possible tests of the final theory, it is clear that any arguments that the final theory is in principle untestable are not valid. In practice it may be hard to do those tests, and we may have to rely on the tests with classes of compactified theories as the best we can do.

PRACTICAL LIMITS?

Another possible limit to our ability to understand nature may arise because society is unwilling to provide the funding and the commitment to do the basic research needed to test the ideas. As I have argued in this book, if there is supersymmetry at the collider scale then there is a chance that the facilities whose upgrade or construction is under way or planned—combined with experiments to study dark matter and proton decay and CP violation and neutrino masses, and with cosmological data—will provide us with the information needed to formulate and test the final theory. Yet even in cases where these facilities are available, there are economic obstacles, because a number of talented and committed people are needed to build and operate these frontier facilities. If the funding comes too slowly, these people will be forced to leave the field, as many had to do when Congress terminated the Superconducting SuperCollider project. The people who do this frontier research can only be based at top research universities and a few national laboratories. If those institutions reduce their commitment, there will be no positions for the theorists and experimenters, or places to train the bright young people who want to learn how and why the universe works, or teachers to train and inspire them.

It is important and interesting to examine the justifications for the funding. In recent years it has been fully understood and documented that much of our economy is based on earlier funding of scientific research. One might think it is the results of research that drive the economy, and they do, but it is not only the results. Fascination with what has been learned from science, and the way science can lead to understanding our world, initially attracts young people to a career in scientific or engineering research. What they end up working on after their education can be very different from what brought them in initially. Many of them see an opportunity to develop products or information technology. The ways that understanding gained from science enriches our culture and our view of our relation to the universe, and the impact of basic research on young people, are probably the two strongest justifications for any society to strongly support basic research.

But even apart from these benefits, basic research more than pays society back for its cost through the mechanism known as "spinoffs." The results of research about Higgs bosons or superpartners or dark matter are not likely to lead to products that affect the economy. But because these research areas are probing new frontiers, scientists must develop new techniques, and these new techniques invariably lead to new industries. The most spectacular example is the World Wide Web, developed at CERN in order to find new ways for international collaborators to handle data from the LEP collider. As Burton Richter likes to say, if they had named it HEP (for high-energy physics) instead of WWW, or if we had a penny for each use of the web, there would be no funding problem in particle physics.

Another example is accelerators, which were invented to probe more deeply into nuclei and protons, and continuously developed to do more particle physics. They now are used to study materials and matter in many ways, providing knowledge about how to make stronger and safer materials, to find the structure of viruses, and much, much more. There are more than thirty thousand accelerators in use in the world today, and less than a few percent are used for particle physics research. Accelerators have so many medical-related uses that the National Institutes of Health have increased funding to support not only the people who use the accelerators but even the accelerators themselves. The list of spinoffs, and of associated start-up companies that can initially survive because of the guaranteed market provided by particle physics labs and experiments, is very long—and it shows convincingly that even if the intellectual interest of the results of the research is not included, funding particle physics and cosmology is an investment that brings large economic returns to society. It is often stated that most new US jobs come from start-ups, and that the majority of start-ups emerge from frontier research. A useful web site for more information on these topics is http://www.interactions.org/cms/?pid=1003378. It would be good if more of the people who control the funding in Washington and at our research universities appreciated this.

ANTHROPIC QUESTIONS AND STRING THEORY

If the laws of nature were different, could life nevertheless exist? Does the universe have to be a certain way for us to be here? More precisely, the particles and forces that determine what happens in the universe have various properties and strengths and ranges—if those were different, what would happen? If Planck's constant or the speed of light were different, would the world be different? These questions were first raised in this modern form in the 1960s and 1970s. It was argued that if any aspect of the laws and constants of nature was any different from what we observed it to be, our universe would be very different and life would not exist. For example, there is an attractive strong force between a neutron and a proton, so that they bind into a deuteron, the second nucleus of the periodic table. From the point of view of the strong force, neutrons and protons essentially behave identically, so there will also be an attractive strong force between two protons. But since the two protons are both electrically charged, they will also feel a repulsive electrical force. In our world that repulsion is sufficient to keep the two protons from binding. However, if the strong force were a little stronger, two protons would bind in spite of the repulsive electrical force. Then the reactions that power the sun would proceed at different rates; the sun would burn its fuel more quickly, and there would not be time for life, dependent on that energy, to evolve on planets. Such arguments are called "anthropic."

Such anthropic questions are clearly not ones that can be answered by experiments, but they are nevertheless research questions that can be addressed from the theory side. If we had the final theory we could work out the answers to such questions. Even assuming only that we will one day have a confirmed M/string theory, we can address a number of anthropic questions. They should be addressed because usually the arguments are oversimplified, because many anthropic questions are addressed by having more complete theories, and also because non-scientific and even religious interpretations are increasingly being given to anthropic arguments whose validity is not established. To give one example, Vaclav Havel, former president of the Czech Republic and a noted writer, has said that "we think the Anthropic Cosmological principle brings us to an idea perhaps as old as humanity itself: that we are not at all just an accidental anomaly, the microscopic caprice of a tiny particle whirling in the endless depths of the universe. Instead we are mysteriously connected to the entire universe."

Anthropic arguments can be split into two kinds. One simply takes account of the fact that people exist, so the universe has to be old enough to allow stars and planets to exist and people to evolve. This is not controversial scientifically—that people exist is data to include as we try to understand the universe. It may be that lots of universes exist, and only some have the properties that allow people to exist. We can call this kind of anthropic explanation "minimal anthropic." There are other kinds of anthropic arguments and explanations that are sometimes discussed; in order to avoid lengthy comparisons we will not examine those here. For our purposes it is sufficient to split anthropic arguments into "minimal" (defined in this paragraph) and "nonminimal" (all the rest). An example of a nonminimal anthropic argument is the claim that human life would not exist if the strength of the strong force had a value slightly different from its actual value, and that the reason it has the value it does cannot be explained scientifically, implying that it has this value so that humans can exist, or even must exist.

One strong reason not to take nonminimal anthropic arguments seriously as implying anything about the world being somehow right for human life is the past existence of the dinosaurs. Earth was a suitable place for them, and they were a dominant species for about 150 million years (nearly three times longer than mammals, and one hundred times longer than humans). Except for a chance asteroid impact 65 million years ago, they would probably still be the dominant species, and mammals (including humans) would not have been able to evolve to their present forms. Any argument about the meaning of the universe should apply equally to the universe 100 million years ago and now. If the universe was planned for humans, somebody got it wrong. Indeed, perhaps one day all humans on Earth will be killed by an asteroid impact—the probability for that to happen is not negligible. If that did happen, would it change how the physical universe and its origin should be explained? The universe would go on, not noticing. Nevertheless, let us examine the nonminimal anthropic issues more technically because they raise interesting questions and suggest issues for the final theory to solve.

Can one conclude that somehow the strengths of the forces are set to their values so that human life will exist? Some people have drawn that conclusion. But most scientists think that such a conclusion is unwarranted by the evidence. First, no one has done the full calculations to determine what actually would happen if the strength of all the forces were different. Often in a complicated physical system, new and subtle interplays at different stages can enter to change the outcome of a calculation. The strength of the strong force enters at a number of places in the full set of processes that fuel the sun. How much can the force change before there is an effect—1 percent? 10 percent? For example, when a major feature of an ecosystem changes, the ecosystem might die, but more often it just readjusts and goes on.

Second, and more important, almost all scientists expect that scientific explanations will be found for all or almost all of the nonminimal anthropic questions. For example, if we did not understand evolution by natural selection, we might think that the way the human body and mind work are evidence of design or planning, an anthropic explanation. But now, after a century and a half of study, extensive documentation has been found to confirm the evolution of the human eye and brain and body. Nonminimal anthropic explanations should not be accepted, nor should any explanation, until alternatives have been explored and found not to do the job. Actually, what kinds of explanations one accepts is a more subtle issue. Nobody can stop someone who prefers to explain nature by saying it was designed for humans to exist. But such an explanation is not scientific, and thus there is no reason for anyone else to accept it. The question is whether there are *scientific* explanations, which can be confirmed by any trained person, are sufficient, and are consistent with the rest of our scientific description of nature. If all nonminimal anthropic issues have scientific explanations, then nonminimal anthropic arguments cannot be interpreted to imply that the universe is a just-so place for humans.

For example, we have seen that in a unified supersymmetric Standard Model, and in M/string theory, the forces of nature are found to be unified. That means that their ratios are fixed by the basic structure of the supersymmetry theory. Then if the strength of the strong force is increased, so must the strength of the electromagnetic force be increased, or the theory would not be a consistent one. If you increase both forces, deuterons are bound more tightly, but protons also repel more strongly, so the behavior of stars may effectively not change. When several quantities vary, each will be able to have a larger range. If that is so, why forces have the values they do is explained scientifically, without reference to whether the universe contains life.

Sometimes proponents of nonminimal anthropic effects try to make the case more compelling by stating it in terms of probabilities. If each force, and several other things, all have some small probability of being in the range needed for human life to occur, and the probabilities are independent, then the probability that they would all be in the required range is the product of the probabilities and therefore very small. When that argument was first made three decades ago, it was worth examining. But now we understand that the forces are unified, so if one force is in the right range the others must also be. The probabilities are not independent, and should not be multiplied, so they are not so small as has often been claimed. Another nonminimal anthropic issue is the need for neutrons to be a little heavier than protons, which translates into requiring up quarks to be lighter than down quarks; those masses are calculable in M/string theory, though they have not yet been calculated. If the world is described by a unique M/string theory, this outcome will not be independent of the others. So even if someone wants to claim that some aspects of the universe we have not yet understood are nonminimal anthropic ones, their probabilities should not be multiplied if they are correlated in the theory, so the combined probabilities would not be nearly so small as is usually claimed.

Until we have the final theory and understand how to calculate its implications we cannot settle all anthropic questions, so they are valid and interesting issues to study. Despite that and the obvious interest of anthropic issues, most physicists do not study them because they expect there will finally be few, if any, anthropic effects that will not be better understood by normal physics. There are, however, two very weak senses in which many more physicists expect anthropic effects to occur. The first one is simply the observation that our universe must have properties consistent with the emergence of human life (i.e., be minimal anthropic). That is simply taking the data into account, and can have no implications for how we give meaning to life and the universe. For example, life will obviously not evolve until stars and planets have formed. In addition, heavy elements are essential for life, and we know that the heavy elements are made in exploding stars (supernovas). So the universe has to be old enough for the first generation of stars to form and grow old and die, and for a second generation of stars to form with planets, before life can evolve.

The second one is more challenging. While we do not yet understand how the universe actually originated, attractive ideas have emerged that provide mechanisms for universes to arise from nothing at all. This is now an active research area. For example, at Stephen Hawking's seventieth birthday celebration, which included a conference, six speakers discussed this topic. Universes might first occur as tiny Planck-scale-size entities, and then inflate in a tiny fraction of a second to sizes large enough to see (were someone there to watch). Their total energy including gravitational attraction is zero, but a large part is in potential energy and that is released in the form of particles—this is what we call the big bang. These ideas imply that new universes are randomly being created by such processes all the time. It may be that the laws of nature are different in different universes. And perhaps the constants that enter into the laws such as force strengths, Planck's constant h, the speed of light c, the gravitational force strength G, and the cosmological constant (really it is only dimensionless ratios of such constants that should be considered, but we will not concern ourselves with that) can be different. Then life may emerge in some universes but not in others.

Even if that is so, it does not imply any anthropic uniqueness for our universe. To understand this, think about a truly random lottery. If there is a big prize and many people enter, the chances that any particular person will win are very small (i.e., the chances that any particular universe will be right for life may be small). However, someone is guaranteed to win (i.e., some universes will have the right properties for life). The person who wins may feel lucky, and thankful, but from the broader perspective we know there is no reason to impute any meaning to which person holds the winning ticket since someone had to win. Similarly, those who end up in one of the universes with life may want to feel some uniqueness, and perhaps even be grateful, but at least from the anthropic point of view that cannot be justified. We don't yet know whether the universe, and the final theory, and the constants and the strengths of the forces are unique. But science is moving toward such understanding.

The Cosmological Constant

If the universe contained some ordinary matter (atoms like those we are made of), and some dark matter, and nothing else, you would expect that the gravitational pull from all that matter would slow down and eventually stop the expansion of the universe. In 1998, physicists compared the expansion rate at different distances from us in the universe. That tells us about the expansion rate at different times after the big bang, because the light takes longer to get to us from farther away. They found the surprising result that the rate of expansion is increasing. There must be additional energy density in the universe that provides an expansion pressure. And there must be a term in the general relativity equations describing the universe that leads to the accelerated expansion. That term is called the cosmological constant (for historical reasons). On the theoretical side, people had tried for a long time to see how big such a term would be if it were there, and always found that the simplest estimates gave huge numbers, so the universe would expand so quickly that no stars or planets could have formed. One can imagine the theory requiring the cosmological constant to be zero by some argument based on symmetries; elsewhere in physics, results like that happen. But it is harder to imagine a calculation getting the actual result, which is tiny compared to the results of naive calculations. There are really two cosmological constant problems: what is wrong with the simplest calculation that gets a very wrong answer, and how could one explain the actual small but nonzero number. These have been called the most important problems in physics and cosmology. We should note that it is not completely settled that the cosmological constant is truly a constant and not a different kind of behavior from a modified theory of gravity or some other mechanism, but the data are within a few percent of a constant value and alternatives are poorly motivated, so we will assume it is a cosmological constant.

One question that arises is whether our inability to understand the cosmological constant problem should prevent us from solving other problems in physics. There is actually no argument that it should affect other results, and the attempts to deal with the cosmological constant problem do not seem to impact other questions. Most theorists think that even if we can describe the rest of our M/string vacuum well it is not likely to help solve the cosmological constant problem, and if the cosmological constant problem gets solved the solution is not likely to help solve any other problems such as Higgs physics, supersymmetry physics, and so on. We will not know for sure until it is solved, of course, but we proceed on the assumption that the cosmological constant problem is essentially orthogonal to achieving the rest of our physics goals. The accelerating expansion of the universe is an important and exciting piece of information, and a challenging one—but to those of us who want to achieve the traditional goal of understanding the world we find ourselves in, it is not the most important problem.

In 1987 (years before the discovery of the accelerating expansion), Steven Weinberg published a paper pointing out that the observed energy density of the universe had about the value it had to have if it were to be a long-lived universe where galaxies and stars and planets and dinosaurs were to exist. He effectively predicted the value of the cosmological constant by requiring that the universe have conditions suitable for the existence of life—it is a fact, a piece of data, that the universe contains life. One way to interpret and understand such a result is to expect that there are many universes and that the value of the cosmological constant is different in different universes. Some of them would have values for the cosmological constant that would allow life to arise, which means the universe is old enough for stars and planets to form.

M/string theory gives a framework in which such a picture could make sense. Recall from Chapter 2 that in quantum field theory one writes a Lagrangian, from which one can calculate predictions for any process, and that Chapter 1 emphasized the important difference between equations and their solutions. The Lagrangian is the equation in quantum field theory that contains all the information about the theory. Any Lagrangian will have many solutions. Any of the solutions can describe physical systems. M/string theory is a framework that replaces the Lagrangian with the 10/11D theory, and there will be many solutions that could describe physical systems. A key idea is that the laws of nature and/or values of particle masses and force strengths, as well as other quantities such as the cosmological constant that determine how physical systems behave, can be different in different solutions. It has been argued that the number of solutions that could arise in the M/string theory approach is huge—so huge that universes with cosmological constants like ours, that allow universes with life, are common. Having the many universes as solutions has been called the "multiverse."

Multiverse ideas are in principle testable, by looking for collisions of universes. The conditions under which they can occur, and the associated signals, are being studied. In addition, if the multiverse is implied in a comprehensive theory that is well verified in a number of ways, then the default would be that it is indeed part of nature. When we have theories, adding or taking away pieces of a theory is dangerous and is likely to lead to misleading consequences. Theorists often write trial effective theories for parts of the world—for example, the Higgs physics sector, or the dark matter. If the world is actually described by an underlying theory such as an M/string theory that encompasses many phenomena, it is easy to write effective theories that are apparently valid but could not be incorporated into the underlying theory. Cumran Vafa coined a nice word for such effective theories—they are in the "swampland."

Always before it had been believed that we could understand physical systems and the underlying laws of nature, or at least we should try until we fail. The new approach to understanding the cosmological constant problem is to give up trying to calculate it—its proponents say the traditional method has failed. Some people are enthusiastic about this generalization of how science works, with one or even some of the things we would normally explain with derivations or calculations being instead explained in a probabilistic sense. Note that for the multiverse explanation to work, it is necessary to have the huge number of possible solutions, so it makes sense only in a framework like the M/string theory one. The existence of a huge number of solutions for M/string theory is a fact, and its relevance to understanding nature needs to be understood. Some reluctantly accept the multiverse solution to the cosmological constant problem as possibly or probably true. Some are angry about having to give up the traditional way science works, and some are trying to find ways to calculate the value of the cosmological constant so that the multiverse is not needed. This is ongoing important discussion and research. Whatever the outcome that emerges, it shows us again that the boundaries of science have changed.

THE ROLE OF EXTRA DIMENSIONS

We saw that the Standard Model, formulated in four dimensions, gave us a descriptive understanding of the particles and forces, but not a "why" understanding. That remained so for the supersymmetric extension of the Standard Model. But compactified M/string theories could address "why" the particles and forces and even the rules are the way they are. Having the extra dimensions, and formulating the underlying theory including their role, may allow us to go beyond a descriptive understanding to a deeper one. The extra dimensions are not a complication but a good thing, an essential aspect of gaining understanding. The value of the Higgs boson mass provides an example. In the Standard Model the value of the Higgs boson mass cannot be calculated at all. In the supersymmetric extension of the Standard Model, the mass can be estimated to be larger than about tens of proton masses and smaller than about two hundred proton masses—better but not what we would hope for. In compactified M/string theories the Higgs boson mass can be predicted with an uncertainty of less than a few percent, just as we would hope would be the case if we had a deep understanding.

Sometimes people comment or argue that the theory has gotten very complicated, with a doubling of the number of particles and with extra dimensions. There is a vague hope that correct theories should be "simple," without fixes and add-ons if problems occur, without epicycles. An "Occam's razor" argument that simpler theories are more likely to be true is sometimes used. Actually, I would argue that the compactified M/string theories are the simplest theories that could possibly encompass and integrate all the phenomena of the natural world into one coherent mathematical theory. The compactified M/string theory that is emerging is as simple as it could be.

THE END OF SCIENCE?

Suppose that eventually we do indeed figure out what the final theory is—and are able to test it so well that most of us are convinced that we indeed understand how our universe works and why it is that way— and the underlying theory framework does (or does not) require a multiverse. What might that imply? Although humans began to wonder about the universe perhaps 40,000 years ago, and started a more systematic quest to understand it about 2,600 years ago in Greece, until about 1600—a mere 400 years ago—there was only a little descriptive progress. When Shakespeare wrote, no aspect of how nature actually works was understood. Then modern science began, and today we have started research on the final theory itself. Before the 1970s we did not know how the world worked, and now, to a large extent, we do. Perhaps in a few decades or less we will reach the end of this quest. Our universe has properties consistent with having humans in it, though it is indifferent to whether they are there, or whether they understand it. For me, and I hope for many people, achieving an understanding of why there is a universe and why it is the way it is will be a source of immense pride and dignity and meaning in the face of that indifference.

Many people, and many scientists, have said that we will not reach that end, that there will always be new questions, that each discovery will lead to additional questions. Why should that be so? There is no known reason why the quest should not end. Consider the analogy with exploration of the surface of the earth: for many centuries there was more to explore, and then one day we were essentially done. There have been a few occasions when someone said that there was nothing new to learn, but if we actually examine the concerns of leading scientists since the 1860s we find that the active scientists always knew that was wrong, and addressed many puzzles. The situation truly is different today. Now we have tested our description back to the beginning of the universe, out to the edge of the universe, and down to the fundamental constituents of matter. Having done so doesn't guarantee that we will achieve total understanding, but it does demonstrate that the historical analogies need not be relevant to the argument.

Yes, it is possible that science in the direction of more fundamental effective theories could end, not because we couldn't get all the way to the final theory but because we did. That wouldn't mean that science itself ended. It would mean we knew the equations and the laws, but we would still be far from knowing all the solutions. There will be more materials and systems to study indefinitely. For complex systems we know that the underlying equations are often of little help in understanding the behavior of the system. As discussed earlier, all areas of science except particle physics and cosmology are open-ended. And if we indeed discover the final theory it will change how we view life and meaning itself.

Appendix

*Predicting the Higgs Boson Mass
from Compactified M-Theory*

In this Appendix I give a short summary of the calculation based on compactified M-theory that correctly predicted the value and approximate properties of the Higgs boson observed at the LHC. I include the Appendix for readers who would like to have a sense of how such calculations work and how M/string theories can be successfully connected to observables and testable physics. Normally in physics predictions depend on assumptions, which is also the case here. The assumptions are expected to be valid, and eventually can be independently checked.

To make any prediction for our world we must first compactify the M/string theory to seven small dimensions near the Planck scale in size, leaving four large space-time dimensions. There are several ways to proceed with this; the theory does not tell us ahead of time which to use. For various reasons we choose to compactify M-theory on a 7D manifold with certain mathematical properties that guarantee the resulting theory is supersymmetric. We look for generic solutions of the compactified theory that behave like our world. Loosely speaking, one can divide all the solutions into two groups: one group has TeV scale interactions (rather than Planck-scale ones), a Higgs mechanism, and other features like those of our world, and a second group does not. We then calculate the ratio of the Higgs boson mass to the measured Z boson mass in solutions from the first group.

The calculation depends on results from a series of papers about the properties of compactified M-theory written between 1995 and 2004 by a number of authors, plus several papers that are part of a six-year program co-led with Professor Bobby Acharya at Kings College, London, and the International Center for Theoretical Physics, and with Piyush Kumar, now at Yale, with results on a number of related topics. This particular calculation of the Higgs boson mass was done by myself with my former student Kumar and two of his present students, Ran Lu and Bob Zheng. There is a recent short review of the whole program, by Kane, Acharya, and Kumar, in the *International Journal of Modern Physics,* vol. A27 (2012).

We make several assumptions, all expected to hold and to eventually be independently checkable. One assumption is that the solution to the cosmological constant problem is separate from the physics of this calculation.

The compactified M-theory prediction of the Higgs boson mass has no free parameters—the compactified theory fixes them all. But today's understanding of string theory is incomplete, and one has to choose the forces and matter content for the compactified theory. Constraints allow only a limited number of choices. We chose the minimal possibility, that the forces are the same as for the Standard Model, and the particles are just the Standard Model ones and their superpartners. It is possible that there are additional forces or matter. For each choice the calculation can be repeated and predictions can be made for the Higgs boson mass, production cross-section, and decay channels. The theory is written near the string and unification scales, not far below the Planck scale. At such scales the electroweak symmetry is not broken. Then one uses standard quantum field theory to calculate the Lagrangian at lower scales, and electroweak symmetry breaking emerges as the TeV scale is approached. One can then calculate the Higgs mass matrix, and the Higgs boson mass is the smallest eigenvalue of the matrix. The Higgs mechanism and the value of the Higgs boson mass are both derived.

The mass predicted by the minimal compactified theory (before the answer was known) was correct, 126 GeV (1.38 times the Z boson mass) ± about 2 GeV. That is an extremely encouraging result for the possibility that compactified string theories can provide a framework that leads to an underlying theory beyond the Standard Model. The crucial part of the result is that certain of the supersymmetry-breaking terms of the Lagrangian are essentially equal to the gravitino mass that sets the whole scale of supersymmetry breaking. This occurs because the gravitino mass can be related to the moduli masses, which in turn have to be larger than about 30 TeV in order to not disrupt nucleosynthesis as the universe cools. This argument is generic for all compactified string theories with moduli, so the result may hold in other corners of string theory too.

The theory makes a number of additional predictions to test it further, and to reduce the number of possible compactification choices. Some are for different decay modes of the Higgs boson, and some for LHC production of gluinos, charginos, and neutralinos. One other related prediction of compactified M-theory was tested very recently. The compactified string theory predicted (already more than three years ago) that a certain rare decay of a meson that is a bound state of a b quark and an anti–strange quark would have a decay to $\mu^+ \mu^-$ that, contrary to a number of other predictions, did not deviate from its expected value in the Standard Model, with no change due to supersymmetry. A combination of theoretical and experimental circumstances make this perhaps odd-sounding example a very good probe of new physics. It was measured by one of the important LHC detectors I have not discussed so much for Higgs bosons and for superpartners, called LHCb. Several other predictions will be tested from LHC data within the next year.

The success of this Higgs mass prediction is not an accident, and provides great encouragement to pursue the string theory framework as an approach to a deeper underlying theory of nature.

Glossary

CONSTANTS

G Newton's constant, which determines the strength of the gravitational force

h Planck's constant, which determines the size of quanta of energy and of quantum theory effects in general

c Einstein's constant, representing the speed of light in vacuum (or charm quark, depending on the context)

POWERS OF TEN

10^{-9} = one billionth, 10^{-6} = one millionth, 10^{-2} = 1/100 (one/hundredth), 10^{0} = 1, 10^{1} = 10, 10^{2} = 100, 10^{3} = 1000, 10^{6} = one million, 10^{9} = one billion

ACRONYMS AND ABBREVIATIONS

Fermilab Fermi National Accelerator Laboratory, Chicago

AMS Detector sensitive to indirect detection of dark matter, on the International Space Station

ATLAS Multipurpose detector at the Large Hadron Collider at CERN

CERN Centre Européan Recherche Nucléaire, Geneva, Switzerland

CMS Multipurpose detector at the Large Hadron Collider at CERN

LEP	Large Electron-Positron Collider, at CERN (turned off so that the LHC could be installed in the same tunnel)
LHC	Large Hadron Collider, proton-proton collider at CERN
LSP	Lightest superpartner
SLC	Stanford Linear Collider, now turned off
TeV	Teraelectron volt, an energy unit, 10^{12} electron volts. A one-volt battery gives an electron one electron volt of energy. Through 2012 the Large Hadron Collider has run with a total energy of 8 TeV and, after a shutdown, will begin running with a total energy of about 13 TeV.

TERMS AND DEFINITIONS

Accelerator

Accelerators are machines that use electric fields to accelerate electrically charged particles (electrons, protons, and their antiparticles) to higher energies. If accelerators are linear, they need to be very long to achieve the desired energies, so some use magnets to bend the particles around and back to the starting point, giving them a little extra energy each time around.

Antiparticle

Every particle has an associated antiparticle, another particle with the same mass but with all charges opposite. If a particle has no charges—for example, the photon—it is its own antiparticle. Often the antiparticle is denoted by writing a bar over the particle name—for example, for the electron antiparticle (also called the positron).

Atom

An atom has a nucleus surrounded by electrons, bound together by the electromagnetic force. Ninety-two different atoms occur naturally, making ninety-two different chemical elements, with nuclei having one to ninety-two protons. The atoms are electrically neutral. The diameter of an atom is about ten thousand times larger than the diameter of its nucleus.

Baryon

A baryon is a composite particle made of three quarks, any three of the six. Protons and neutrons are baryons.

Baryon Asymmetry. *See Matter Asymmetry*

Beams

One way to learn more about particles is to collide them with one another and see what happens. Beams of electrons and protons can be made by knocking apart hydrogen atoms and applying electric fields. Positrons and antiprotons don't exist naturally, because they annihilate as soon as they encounter an electron or proton; they can be made by hitting a target with energetic protons or electrons, and then collected by placing magnets after the target, arranged so as to bend each kind of particle in a different path. Then bunches of them are accelerated to higher energies. When a particle hits a target, every kind of particle is made with a certain probability, so other beams of particles can also be made (neutrons, muons, kaons, neutrinos, etc.) by judicious arrangements of magnets and material.

Big Bang

Several strong kinds of evidence imply that our universe began as a tiny dense gas of particles that has been expanding since—in other words, that it began in a "hot big bang." One argument is that the universe is indeed expanding. A second is that the abundance of light nuclei such as helium was correctly predicted, and a third is that the temperature and frequency spectrum of the electromagnetic radiation today, the cosmic microwave background radiation, was correctly predicted.

Bind

When particles feel electromagnetic or QCD attractive forces, they will form a bound system. Often it is said that they "bind into" an atom or a hadron or some composite system. The proton is a bound system of three quarks, with the QCD or strong force carried by gluons, and the hydrogen atom is a bound system of an electron and a proton with the electromagnetic force carried by photons.

Black Hole

Black holes form whenever sufficient matter is packed into a small enough space so that the resulting gravitational force at the surface is strong enough to prevent anything, including light, from escaping. They can be microscopic in size, or formed from billions of stars. They can be detected from their indirect effects on nearby matter. The theoretical study of their properties can greatly clarify our understanding of basic questions.

Boson

Bosons are any particles that carry an integer unit of spin (0, 1, . . .). Their properties are different from those of particles with half a unit of spin (fermions). In particle physics, "boson" has a more specific usage: bosons (photons, gluons, Ws, and Zs) are particles that are the quanta of the electromagnetic, strong, and weak fields. They transmit the effects of the forces between quarks, leptons, and themselves. Higgs bosons are quanta of a Higgs field.

Charge: Electric, Color, Weak

Each particle can carry several kinds of charge that determine how it interacts with others. Electric charge is familiar to us in everyday life. Particles can have positive or negative electric charge, or none. Color charge and weak charge are not familiar because their effects can be felt only at distances smaller than the size of a nucleus. Color charge and strong charge are the same thing. A particle cannot have random amounts of charge; only certain discrete amounts are allowed. The extent to which a particle feels each force is proportional to its associated charge. Quarks and gluons carry color charge; quarks and leptons, and W and Z bosons, carry weak charge.

Chemical Elements

Ninety-two different stable nuclei can be formed from neutrons and protons bound together. Each forms atoms by binding as many electrons to the nucleus as it has protons (so the atom is electrically neutral), giving ninety-two different atoms. These atoms are the smallest recognizable units of the ninety-two chemical elements.

Cold Dark Matter

Particle physics theories that extend the Standard Model often predict the existence of new, stable particles that were present in the early universe and survive to the present, making up a large fraction of the matter of the universe. These particles interact weakly, and they are usually massive so they move slowly—they are cold. An example of such a particle is the lightest supersymmetric partner. Astronomers have evidence from the motion of galaxies, and from the large-scale structure of the universe, that cold dark matter exists. There are several candidates that each could be part of the cold dark matter.

Collider

A collider is made by accelerating beams of particles and causing them to collide. The energy of the colliding beams can provide much more energy that can be used to make new particles than if the beams hit stationary targets. Colliders have two challenges: getting to larger energies, and getting to higher intensities.

Color Charge. See Charge: Electric, Color, Weak
Color charge means the same thing as strong charge.

Color Field

Any particle carrying color or strong charge has an associated color or strong field around it. Any other particle carrying color charge feels that field and interacts with the first particle.

Color Force

The force between two particles carrying color charge. The color force (or strong force) binds quarks into protons and neutrons. The residual color force outside protons and neutrons is the nuclear force that binds protons and neutrons into nuclei. The color force is mediated by the exchange of gluons.

Compactify

For a ten- or eleven-dimensional theory to explain or describe our world, the extra dimensions have to become invisible to us. One way to do that is to have them become very small, Planck-scale size. The process of taking the full theory and writing it as a combination of the theory of our four-dimensional space-time plus the curled-up small dimensions is called "compactifying."

Composite

Any object made of other objects is composite, as are atoms, nuclei, and protons. If quarks and leptons had followed the historical trend that each level of matter turned out to be composites made of smaller constituents, experiments should already have shown evidence of their compositeness. That, combined with theoretical arguments, strongly suggests quarks and leptons may be the ultimate constituents of matter, the indivisible "atoms" of the Greeks.

Constituents

Any objects that are bound together to make larger objects. For example, atoms are constituents of molecules, nuclei are constituents of atoms, and so on. (*See also* Composite.)

Cosmic Rays

Protons and some nuclei that are ejected from stars, especially supernova explosions, move throughout all space. They impinge on the earth from all directions and are called "cosmic rays." They normally collide with nuclei of atoms in the atmosphere, producing more "secondary" particles, mainly electrons, muons, pions, and so on. A number of cosmic-ray particles go through each of us every second, and they can interact in detectors and mimic signals of previously undetected particles, so experimental equipment must shield against them or be able to recognize them so they can be discounted as signals of new physics.

Cosmological Constant

The cosmological constant is the name of a possible term in the equations that describe the universe. If it is not zero, then it implies there is a force that is slowly increasing the expansion rate of the universe. There are two puzzles about the cosmological constant. First, the observed value seems to be far smaller than estimates would imply, and, second, recent data suggest it is not exactly zero, so an explanation is needed for why it has a particular nonzero value.

Cosmology

Cosmology is the study of the universe as a whole, its properties, and its origin.

CP Violation

Interactions of quarks, leptons, and bosons are normally invariant under a symmetry operation called "CP," the combined operations of "Parity" and "Charge Conjugation." A small violation of this invariance is observed, which may have important implications and be an important clue to deeper understanding of how nature works.

Dark Matter

Particle physics theories that extend the Standard Theory predict several forms of matter that may exist in large quantities throughout the universe and make up most of the matter of the universe. Some move slowly and are called "cold dark matter"; others move rapidly and are called "hot dark matter." Study of the motions of galaxies and of the formation of clusters of galaxies suggest that such dark matter exists, as do theoretical cosmological arguments based on other data. (*See also* Cold Dark Matter.)

Decay

The quarks and leptons and bosons that are the particles of the Standard Theory have interactions that allow them to make transitions into one another. Whenever one of them can make transitions into lighter ones, that transition will occur with a certain probability, and we say the heavier one is unstable and has decayed into the lighter ones. Decays are really transitions—the final particles were not contained in the initial one. The initial particle disappears, and the final ones are created. In the Standard Model the up quark, the electron, and the neutrinos do not decay; the other fermions, and the W and Z, do decay. All of the superpartners except the lightest one are expected to decay.

Detector

Physicists study the properties of particles and their interactions by observing their interactions and decays. These observations are done with detectors, which can be thought of as cameras that record information in several ways, not just on film. In order to obtain new results at the forefront of today's questions, detectors have to be very large and use (and also often require development of) better technologies. Every particle physics experiment has one or more detectors.

Deuteron

A deuteron is the second heaviest nucleus (after the lightest, hydrogen, which has a single proton). It is composed of a neutron and a proton bound together by the nuclear force. The deuterium atom has a single electron bound to a deuteron (one electron because there is one proton).

Dimensions

Mathematically one can generalize equations to describe worlds with different numbers of dimensions. For a century physicists have been doing that, and studying the possibility that our world has more than four space-time dimensions, some perhaps curled up to be too small to simply observe. String theories must have ten space-time dimensions to be mathematically consistent.

Dirac Equation

The Dirac Equation incorporates the requirements of both quantum theory and special relativity in the description of the behavior of fermions. It requires that fermions have the property called spin, and it predicts the existence of antiparticles. It was written by Paul Dirac in 1928.

Effective Theory

Each part of the physical world can be described by a sub-theory that applies over a certain distance or energy scale. Such sub-theories are called effective theories. Explanations in a given effective theory can ignore much of the rest of the world. The rest has effects on the part of interest through a few inputs or parameters. Every part of our description of the physical world is an effective theory except the ultimate theory, which in this book is called the final theory. (*See also* Chapter 3.)

Electric Charge. See Charge: Electric, Color, Weak

Electromagnetic Force. See Electroweak Force

Electron

A fundamental particle. The electron has one unit of negative electric charge and one-half unit of spin. It is a fermion.

Electroweak Force

The descriptions of the electromagnetic and weak forces have been unified into a single description, the electroweak force. The electromagnetic and weak forces appear to be different because the W and Z bosons that mediate the weak force are massive, while the photon that mediates the electromagnetic force has no mass. Consequently, it is easier to emit photons than W and Z bosons. The electroweak unified theoretical description treats all the bosons on an equal footing—and explains why they appear to be different—because of their masses.

Family

Quarks, and leptons, appear to come in three families, even though only one family appears to be needed to explain the world we see. The other families differ only in that they are heavier. We do not yet understand why there are three families, but we can fully describe their behavior. This is one of the main mysteries of particle physics. It is addressed by string theory but we don't know the actual solution yet.

Fermion

Fermions are particles with half a unit of spin. Their properties are different from those of particles with an integer unit of spin (bosons). Quarks and leptons, the matter particles, are fermions.

Feynman Diagrams

The rules of any quantum field theory can be formulated so that it is possible to draw a set of (Feynman) diagrams representing the processes that can occur, and to assign a probability of occurrence to the process represented by each diagram or set of diagrams. The diagrams are very helpful as guides to thinking about what processes can occur. The structure of the theory determines which diagrams are allowed.

Field

Every particle is the origin of a number of fields, one for each nonzero charge the particle carries. Interactions occur when one particle feels the field of another (and vice versa). There are electromagnetic fields, weak fields, and strong fields. Any particle with energy (remember, mass is a form of energy) sets up a gravitational field. In the Standard Model, particles get mass by interacting with a Higgs field, but the origin of the Higgs field is not from a charge.

Final Theory

The name used in this book for the theory sought by many particle physicists that includes not only the Standard Theory but also the theory of gravity, and explains why the final theory itself takes the form it does, and explains what quarks and other particles are, and explains what space and time are, and more.

Forbidden

Processes can be naively imagined that might occur, but should not occur according to the predictions of the Standard Model. Whether they occur is then a test of the Standard Model. If they occur at the same rate as other processes, the Standard Model would be wrong; if they occur at much smaller rates, or do not occur at all, they provide a clue as to how to extend the Standard Model. None of the processes forbidden by the Standard Model have been observed.

Force

All the phenomena we know of in nature can be described by five forces: the gravitational, weak, electric, magnetic, and strong forces. The electric and magnetic forces are unified into the electromagnetic force. Although the weak and electromagnetic forces appear different to us, they can be described as unified into one force (electroweak) in a more basic way; there is evidence that a similar unification of the electroweak force with the strong force also occurs. The attempt to unify all forces is an active research area. In particle physics, "force" and "interaction" mean essentially the same thing.

Gauge Boson

The strong, electromagnetic, and weak interactions are transmitted by the exchange of particles called gauge bosons (gluons, photons, and Ws and Zs). The gauge bosons are the quanta of the strong, electromagnetic, and weak fields.

Gauge Theory

A gauge theory is a quantum field theory where interactions occur between particles carrying charges, with strengths proportional to the sizes of the charges, and are transmitted by bosons that are quanta of the fields set up by the charges. In a gauge theory, once any particle exists (such as an electron) that carries any kind of charge (electric or weak or strong or any yet to be found), the associated boson that transmits the force must exist (photons or W and Z bosons or gluons or any yet to be found), or the theory would not be consistent.

General Relativity

Einstein's theory of the gravitational interaction.

Generic

"Generic" refers to results that are obtained without any special choices of parameters or situations or conditions or constraints. They may not be theorems but they hold very generally. One would have to work hard to find or construct exceptions (if that is possible at all), and to show that possible exceptions are not already excluded by some data or constraint.

Gluino

The hypothetical supersymmetric partner of the gluon, differing only in that the gluino has spin one-half while the gluon has spin one, and the gluino is heavier.

Gluon

The particle that transmits the strong, or color, force; the quantum of the strong field.

Gluon Jet

The color, or strong, force is so strong that colored particles (quarks and gluons) hit or produced in a collision can separate from other particles carrying color charge only by binding to one another, and by making color-neutral particles (hadrons), mainly pions. Thus an energetic gluon or quark becomes a narrow "jet" of hadrons as it moves along, turning its energy into the mass and motion of several hadrons. A quark or gluon appears in a detector as a jet of typically five to fifteen hadrons.

Grand Unification

The proposed unification of the weak, electromagnetic, and strong forces into a single force. This unification, if it occurs, must happen in the sense that the forces act as a single one at very short distances, a million billion times smaller than distances that have been studied experimentally so far; the forces behave differently when they are studied at larger distances.

Gravitational Force. See Force

Gravitino

The superpartner of the graviton. When supersymmetry is broken the gravitino become massive, and the splitting of the gravitino and graviton masses sets the scale of all the superpartner masses. Since the graviton remains massless, the gravitino mass is the basic mass scale of the broken supersymmetric theory.

Graviton

The quantum of the gravitational field, which mediates the gravitational force.

Hadron

The properties of the color force and the rules of quantum theory allow certain combinations of quarks (and antiquarks) and gluons to bind together to make a composite particle; all such particles are called hadrons. When mainly three quarks bind, the resulting hadron is called a "baryon." When quark and antiquark combine, the result is called a "meson," and when gluons combine it is called a "glueball." Hadrons have diameters of about 10^{-13} cm. The proton and neutron are the most familiar baryons. Pions are the lightest mesons so they are produced frequently in collisions. Kaons are the next lightest hadrons; their properties make them useful in many studies.

Helium Abundance

As the universe cooled after the big bang, it eventually reached a stage (about a minute after the beginning) when protons and neutrons formed, and then nuclei. Nuclei up to helium formed, but collisions were too soft for heavier nuclei to form. The theory of the big bang predicts the fraction of nuclei that are helium, and that fraction has been measured; the observed amount agrees very well with the predicted amount. This is one of the main reasons that the universe is generally believed to have begun in a hot big bang.

Higgs Boson. See Chapters 2 and 7.

Higgs Field

In the Standard Model some particles (bosons and fermions) get mass by interacting with the Higgs field. The Higgs field and the way the particles interact with it must have very special properties for the masses to be included in the theory in a consistent way. The other fields we know of arise from particles that carry charges, but the Higgs field does not. (*See also* Higgs Mechanism.)

Higgsino

The superpartner of the Higgs boson.

Higgs Mechanism

The Higgs mechanism is a special set of circumstances that must hold if bosons and fermions are to get masses from interacting with a Higgs field, even if the Higgs field exists. In the Standard Model these circumstances can be imposed, and in the supersymmetric Standard Model they can be derived.

Higgs Physics

This is the combined physics that explains the origin of the Higgs field, the reason the Higgs mechanism applies, and the properties and study of the Higgs bosons.

High-Energy Physics

Another name for particle physics, often used because much of particle physics (though not all) is based on experiments requiring high-energy beams.

Hot Dark Matter. See Dark Matter

Inflationary Universe

According to the inflationary universe theory, the universe went through a stage of very rapid expansion, called "inflation," and then slowed down to the present expansion rate. The inflation was driven by a large energy density that was unstable and decayed by releasing its energy into particles with kinetic energy.

Intensity

A measure of how often collisions occur at a collider. (*See also* Luminosity.)

Interaction. See Force

Jet. See Gluon Jet

Kaon. See Hadron

Lagrangian

A Lagrangian is an equation that contains representations of all of the fundamental particles in the world and specifies how they interact. Given the Lagrangian, the rules of quantum theory specify how to calculate the behavior of the particles, how to build up all of the composite systems they form, and what all of the consequences of the basic theory are.

Lepton

A class of particles defined by certain properties: leptons are fermions with spin one-half that do not carry color charge and that have another property called lepton number that is different for each family. The known leptons are the electron, the muon, the tau, and their associated neutrinos.

Lightest Superpartner (LSP)

The superpartner with the least mass may have several important roles. In particular, it may be the cold dark matter of the universe, and its properties are crucial for identifying the events of superpartner production at colliders, since all of the heavier superpartners decay into the lightest one.

Luminosity

Any collider has two basic figures of merit: the maximum energy it can supply to the collisions, and how often it can cause collisions to occur. The number of events at a collider over some period of time is the product of two factors: the probability that if two particles actually collide something will happen, and the number of collisions. The latter is just a property of the collider, not of the physics that governs the collision. It is called the luminosity. It depends on features such as how many particles can be accelerated, how tightly bunches of them can be packed, and so forth. Loosely speaking, we can refer to the luminosity as the intensity.

Mass

Mass is an intrinsic property of any object; it measures how hard it is to make the object move. It can be thought of as weight, though the two are not quite the same (the mass of an object does not change, but if it were transported to a planet with a different mass its weight would change). From the particle physics point of view, all mass is thought to arise from interactions.

Matter

It is useful to think of quarks and leptons as the basic particles that make up all the things around us, and the photons and gluons that bind them as quanta of the fields. We call the quarks and leptons matter particles. Sometimes the term "matter particles" is used to mean fermions.

Matter Asymmetry

Our universe seems to be made of matter, but not antimatter, so there is an "asymmetry." Several ideas exist to explain how a universe could initially be symmetric, with equal numbers of protons and antiprotons, but evolve into today's asymmetric one, with about a billion protons for every antiproton. This is an active research area. (Matter asymmetry is sometimes referred to as baryon asymmetry.)

Maxwell's Equations

Electromagnetism, the unified theory of all electric and magnetic phenomena, is summarized in a set of equations, first written by Maxwell in the 1860s. When they are extended to include the effects of the quantum theory, the theory of quantum electrodynamics (QED) is obtained. Physics students spend about a quarter of their time for two years learning how to solve Maxwell's equations, unless they plan to work in a subfield that relies heavily on Maxwell's equations, in which case they spend much more time studying them.

Mediate

The effects of interactions are transmitted from one particle to another by exchange of particles called bosons. The bosons are said to mediate the interaction or force.

Meson. See Hadron

Microwave Background Radiation

As the universe expanded and cooled, the original particles decayed or annihilated until only photons, neutrinos, protons, neutrons, and electrons that formed atoms remained. Today there is a cold gas of photons, about four hundred in each cubic centimeter of the universe, called the microwave background radiation, because the wavelength of the photons is in the microwave part of the spectrum. The properties of this background radiation can tell us a great deal about the properties of the universe and how it began, and are the subject of intense study.

Missing Energy

When superpartners are produced at colliders, we expect each superpartner to decay into Standard Model particles plus the lightest superpartner, which interacts weakly and thus escapes the detector. The energy it carries off is expected to be one of the main signatures that tells us superpartners have been produced.

Moduli

When a theory is compactified so that it has three large space dimensions and seven or six small ones, it is necessary to have information about the sizes and shape of the small dimensions in order to calculate predictions of the theory. For example, one needs the diameters along the various dimensions, the overall volume, and orientations of planes formed by two dimensions compared to others. As always in physics, such information is described by quantum fields. These fields are called "moduli." There is nothing like them in our everyday experience, so it is hard to explain and describe them. Both the fields and their quanta are called "moduli"; the distinction becomes clear from the context. To make definite predictions, the moduli must take on definite values, which can be different in different vacua. When the theory leads to definite values for the moduli, it is referred to as "stabilizing" them. The moduli quanta are unstable and will always decay via their gravitational interaction with Standard Model particles and superpartners.

Molecules

Although atoms are electrically neutral, the positive and negative charges are not on top of one another so there is some electric field outside an atom. Therefore, atoms can attract each other, and form molecules, which can get very large.

M/String Theory. See String Theory

M-Theory
Compactified M-theory is a candidate to describe our string vacuum.

Muon
A fundamental particle. A muon decays into an electron and neutrinos in about a millionth of a second. Muons are made in collisions at accelerators, and in decays of other particles produced at accelerators and in cosmic-ray collisions.

National Laboratories
Since much of the research in particle physics has to be done at large accelerators which are very expensive, the accelerators are built as national or international facilities at a few labs and used by many particle physicists.

Neutrino
A fundamental particle. There is one neutrino for each of the three families of particles.

Neutron. See Hadron
Free neutrons have a lifetime of about fifteen minutes; they decay into a proton and an electron and an antineutrino. When the neutrons are bound into nuclei (such as those in us), the decays are no longer possible because of subtle effects explained by quantum theory, so the neutrons in nuclei are as stable as protons.

Newton's Constant G
Newton's law of gravitation says that the gravitational force between any two bodies is proportional to the product of their masses and

decreases as the square of the distance between them. This statement is turned into an equation by inserting the constant G, so the force $F = Gmm'/r^2$. Since G is universal because all of the particles feel the gravitational force, G can be used to form quantities with dimensions, giving the Planck scale. G is measured by finding the force between two objects of known masses separated by a known distance.

Newton's Laws

Newton formulated the law that describes the gravitational force (*see* Newton's Constant G) and three laws that describe motion. The first law says that every moving body moves in a straight line at constant speed unless a force acts on it; the second law says that the product of the mass of a body and its acceleration is equal to the force acting on it ($F = ma$); and the third law says that if one body applies a force on a second, then the second applies an equal and oppositely directed force on the first.

Nuclear Force

Although protons and neutrons carry no strong or color charge, at tiny distances near a proton or neutron the cancellation of the strong field from its constituent quarks and gluons is incomplete, leaving a residual strong force that leaks outside the proton and neutron. This is the nuclear force that binds protons and neutrons into nuclei.

Nucleus

Although protons and neutrons are color-neutral composites of quarks and gluons, the quarks and gluons are not all at the same places so some of their color, or strong, fields exist outside the proton or neutron, giving an attractive force that binds protons and neutrons into nuclei. The attractive effects of this residual color force is offset by the electrical repulsion of the protons, so nuclei with too many protons cannot exist. It turns out that there are ninety-two stable or long-lived nuclei in nature. They are the nuclei of the atoms of the ninety-two chemical elements.

Particle

The term "particle" is used somewhat loosely, and includes not only the elementary quarks and leptons and bosons but also the composite hadrons. It also includes any (currently hypothetical) new particles that might be discovered, such as the supersymmetric partners of the quarks and leptons and bosons.

Particle Physics

This is the field of physics that studies the particles and tries to understand their behavior and properties. Sometimes a distinction is made between particle physics that studies quarks, leptons, gauge bosons, and Higgs physics, on the one hand, and the study of hadron physics that aims to relate the properties of the hadrons to the theory of the color force, on the other. More broadly, the goals of particle physics are to understand not only the description of the particles and their interactions but also why the laws of nature are what they are, and how the universe arises from those laws.

Photino

The supersymmetric partner of the photon.

Photon

The photon is the particle that makes up light. It transmits the electromagnetic force. It is the gauge boson of electromagnetism. Once electrons exist, quantum theory implies that photons must exist and have the properties they do.

Pion

The lightest hadron, and therefore the one most often produced in collisions. (*See also* Hadron.)

Planck Energy, Length, Time. See Planck Scale

Planck Scale

Planck scale refers to certain values of length, time, and energy or mass. To understand how these values originate, suppose you were trying to explain to an intelligent being in another galaxy how long humans typically lived. You couldn't use hours or years since those units are defined on Earth (for example, by how long it happens to take our planet to go around its sun once), so a being in another galaxy wouldn't know what you meant. However, every scientist in the universe knows the values of Planck's constant (h), the speed of light in vacuum (c), and the universal strength of the gravitational force (G). You could use those to form ratios that define a universal unit of time called the Planck time, and tell the being from another galaxy our typical lifetime in units of Planck times. Similar units for length and mass or energy can be defined. Max Planck realized this possibility and defined these units at the beginning of the twentieth century. Since the Planck scale units are the only universal ones, we expect the fundamental laws of nature to be simple in form when expressed in those units. (*See also* Chapter 3.)

Planck's Constant h

Many things are quantized, such as the energy levels of atoms. Planck's constant sets the scale of quantization: energy levels are separated by amounts proportional to h, the amount of spin a particle can have is a multiple of h, and so on. Planck originally found that the energy radiated by a heated body was emitted in quanta, rather than emission of any continuous amount being possible. The amount of energy was always an integer multiple of hf, where f is the frequency or color of the radiation; that is, hf of energy can be emitted, or $2hf$, or $3hf$ and so on, but not amounts in between. By separately measuring the frequency, Planck deduced the value of h, $h = 6.63 \times 10^{-34}$ Joule-seconds. (*See also* Planck Scale.)

Point-Like

If matter is probed with projectiles that are large, and that have energies that are less than what is needed to change the energy levels of an atom, then atoms will seem to be point-like objects. If the energy is increased, eventually the projectile will penetrate the atom but encounter the nucleus that will seem to be point-like. With higher energy the nucleus will appear to be made of point-like protons and neutrons. With still higher energies the protons and neutrons will be seen to be made of point-like quarks and gluons. As the energies of projectiles were increased still more, quarks and leptons might have been seen to be made of something still smaller, but that has not happened: they behave as point-like up to the highest energies they have been probed with, energies well beyond those for which we would have expected to find more constituents if history were to repeat itself once more. Further, the structure of the Standard Model theory suggests that quarks and leptons are the fundamental, point-like constituents of matter.

Positron
The antiparticle of the electron.

Predict
The term "predict" is used in the normal sense that a theory may predict some unanticipated or as-yet-unmeasured result. It is also used in another sense: a theory can be said to predict a result that is already known, because once the theory is written it gives a unique statement about that result. Sometimes an in-between situation holds, in that the theory predicts a result uniquely in principle, but the prediction depends on knowing some other quantity or requires very difficult calculations.

Projectile
One way to study particles and their interactions is to probe them with projectiles. The projectiles are other particles—electrons, photons, neutrinos, and protons—because these are small enough and can be given enough energy.

Proton. See Hadron

Proton Decay

If the Standard Model were the complete theory that described nature, protons would be stable, never decaying. If the Standard Model is part of a more comprehensive theory that unifies quarks and leptons, then possibly protons are unstable, though with extremely long lifetimes. Some basic theories imply that protons decay, while others do not. Experiments that search for proton decay are very important, because if we knew it occurred (and what the proton decayed into), it would provide valuable information about how to extend the Standard Model.

Quanta

Each particle is surrounded by a field for each of the kinds of charges it carries, such as an electromagnetic field if it has electrical charge. In the quantum theory the field is described as made up of particles that are the quanta of the field. More loosely, quanta refers to the smallest amount of something that can exist.

Quantum Field Theory

When interactions among particles are described as transmitted by exchange of bosons, the methods of quantum field theory are used.

Quantum Theory

The quantum theory provides the rules to calculate how matter behaves. Once scientists specify what system they want to describe, and what the interactions among the particles of the system are, then the equations of the quantum theory are solved to learn the properties of the system.

Quark

A fundamental particle. Quarks are very much like electrons, but also carry strong charge and thus have another interaction, one that can bind them into protons and neutrons. There are six quarks, called up (u), down (d), charm (c), strange (s), top (t), and bottom (b).

Quark Jet

Because quarks must end up in hadrons, quarks that are produced in collisions actually appear in detectors as a narrow jet of hadrons, mostly pions. (*See also* Gluon Jet.)

Radioactive Decay

Some nuclei are unstable, but live long enough to exist as matter until they decay. When they decay they can emit several particles—photons, electrons, positrons, neutrinos, neutrons, and even helium nuclei. For historical reasons such decays were called "radioactive decays." Sometimes scientists use the emitted particles as tools to do experiments.

Reductionist

One way to study the natural universe is to study very detailed aspects of nature, to take things apart and see what they are made of, and to focus on small steps. This approach is called "reductionist." It has been a powerful success, letting us build up the remarkably complete description of nature we now have. Whenever possible, scientists have tried to unify subfields as they became understood. Recently in particle physics the trend toward unification has been increasingly successful. For physicists, reductionism includes the associated unification. (*See also* Unification.)

Relativistic, Relativistic Invariance

Whenever particles can move at speeds near the speed of light, and whenever fields are involved, the description of nature must satisfy the requirements of Einstein's "special relativity" theory.

Rules

In order to have a complete understanding of nature, we need to know the particles, the forces that determine the interactions of the particles, and the rules for calculating how the particles behave. For the motion of objects normally on Earth or in the sky, the rule to use to calculate the behavior of particles is Newton's second law, $F = ma$. When atomic distances or smaller are involved, the Schrödinger equation of quantum theory replaces Newton's second law. In particle physics, additional relativistic requirements are added to make the complete set of rules— namely, quantum theory and Einstein's special relativity. (*See also* Chapter 1.)

Schrödinger Equation

The equation from quantum theory that tells how to calculate the effects of the forces on the particles. It is the quantum theory equivalent of Newton's second law.

Science

Science can be defined as a self-correcting way to get knowledge about the natural universe, plus the body of knowledge obtained that way. It is both a method and the resulting understanding and knowledge. The method requires making models to explain phenomena, testing them experimentally, and revising them until they work. The goal of science is understanding. Once part of the natural world is understood, it may be possible to develop applications of the new knowledge. The process of developing such applications is properly called technology, not science. Although scientific knowledge may, and usually does, lead to technology, science is not necessary for technology, and technological developments have led to new science as much as the opposite. Before the time of Galileo, many technological developments occurred that had no scientific connection. Since the time of Maxwell and his writing of the electromagnetic theory, almost all technological developments have depended on earlier science. In recent years the words "science" and "technology" have been frequently misused, as if they were interchangeable. Because science and technology are really different, it is better to carefully distinguish between them.

Selectron

The supersymmetric partner of the electron.

Signature

A new particle will have some characteristic behavior in a detector that allows it to be recognized. Particles that decay into others do so in a unique way that is different for every particle. Knowing the properties of the particle allows us to calculate how it will decay. The features that allow a new particle to be identified in a detector are called its signature.

Slepton

The supersymmetric partner of any of the leptons.

Smatter

The superpartners of the Standard Model particles. This book argues that the experimental discovery of smatter will provide us with information that will be essential for gaining insights into the ultimate laws of nature, the final theory.

Solar Neutrinos

The reactions that fuel the sun lead to the emission of photons that reach the earth as sunlight, and of neutrinos that we do not see with our eyes, but which can be detected in special neutrino detectors.

Special Relativity

The constraints of special relativity are two conditions that Einstein pointed out should be satisfied by any acceptable physical theory. Somewhat oversimplified, these conditions are, first, that light moves at the same speed in vacuum regardless of how it is emitted, and, second, that scientists working in different labs moving with different relative speeds should formulate the same natural laws. The constraints imposed by these conditions have surprising implications for the structure of acceptable theories. For example, the Schrödinger equation of quantum theory does not satisfy these conditions. But when it was generalized by Dirac to do so, the resulting equation led to the prediction of antiparticles, which need not have existed from the point of view of quantum theory alone.

Spectra

Atoms can exist in a number of discreet energy levels. They emit or absorb photons when they make transitions from one level to another. The energies of the photons emitted or absorbed by one atom are different from those of all other atoms. The photon energies are directly related to their frequencies, which set their colors in the spectrum, so by observing the colors of the photons we can determine which atoms are being observed. This can be done in a laboratory, and it can also be done with the light reaching us from stars, near or distant, which allows us to identify the atoms that stars are made of. Only the same ninety-two elements we find on Earth are seen throughout the universe.

Speed of Light

Light and all other massless particles travel in vacuum with a speed, usually labeled c, whose value is about 300 million meters a second. Special relativity implies that no particle or signal can move faster than the speed of light, and that photons always have this speed regardless of the speed of their source.

Spin

Spin is a property that all particles have. It is as if particles were always spinning at a fixed rate (which could be zero), which can be different according to the type of particle. It is not quite right to think of them actually spinning because the particles do not have to have spatial extension to have spin; calling this property "spin" is an analogy. The amount of spin is required by the quantum theory to come in definite amounts; if the unit is chosen to be Planck's constant h divided by 2π, then particles can have zero spin, half a unit of spin, one unit of spin, and so on.

Spontaneous Symmetry Breaking

Often the equations of a theory may have certain symmetries, though their solutions may not; the symmetries are hidden, or broken. For example, the equations may describe several particles in identical ways, so the equations are unchanged if the particles are interchanged, but the solutions may give the particles different properties (as illustrated in a simple example in Chapter 1). This phenomenon is called spontaneous symmetry breaking.

Squark
The supersymmetric partner of any of the quarks.

Stable Particle
Some particles do not decay into others; they are called stable. (*See also* Decay.)

Standard Model
The very successful theory of quarks and leptons and their interactions that is described in this book is called the "Standard Model" by particle physicists. The name arose historically as the theory developed, and then was difficult to change because it is widely used. The Standard Model is the most complete mathematical theory of the natural world ever developed and is well tested experimentally.

String Theory
String theory, or M-theory, is a framework that can provide the ingredients for a full theory of rules, forces, and particles. In this book I refer to M/string theory. Chapter 8 provides explanations. A consistent mathematical theory can only exist with extra space dimensions beyond our familiar three dimensions. In this book I consider only small extra dimensions, essentially Planck-scale size. In order to explain our world, the ground state or vacuum of the full theory, the ten- or eleven-dimensional theory must be "compactified" to the familiar three space dimensions. Somewhat oversimplified, string theory is a theory that aims to unify all of the forces and particles of nature and explain why they are as they are. In string theory there is only one force (gravity) in ten dimensions, but when looked at from our four-dimensional world, the extra dimensions imply the other forces we observe. Particles are

strings that vibrate in different ways to account for their various properties. String theory appears to allow the construction of a quantum theory of gravity.

Strong Force. See Force

Structure

Objects have structure if they have parts—if they are made of other things. Whether objects have structure can be learned from experiments that probe them with projectiles. Over the past century each stage of matter that was found, as it became possible to search for smaller things, turned out to have structure. Quarks and leptons appear not to have structure, so perhaps the search for the basic constituents has finally ended. There are also theoretical arguments that quarks and leptons are the basic constituents.

Subatomic Particle

Any particle that is contained in an atom, or any particle that can be created in collisions of such particles, is loosely called "subatomic," whether it is composite like a proton or elementary like a quark or electron.

Superpartner

If the theory that describes nature has a symmetry called supersymmetry, then every normal particle (the ones we know) has associated with it a partner that differs only in its spin and its mass.

Superspace

Supersymmetry can be formulated in several ways. One is to imagine associating another coordinate having special properties with each of our normal space-time coordinates, giving a kind of space called superspace. Writing theories in superspace makes them supersymmetric. This way of constructing supersymmetric theories is harder to picture than associating superpartners with each Standard Model particle, but it leads to the same results and sometimes facilitates deriving mathematical properties of the theories.

Superstring

String theories are expected to be supersymmetric, and are often called superstring theories.

Supersymmetry

A hypothetical symmetry that describes nature. It says that even though fermions and bosons seem to us to be very different in their properties and their roles, in the theory itself they appear in a symmetric way. If supersymmetry is indeed realized in nature, then every particle has a superpartner.

SUSY

This is a common abbreviation for supersymmetry.

Technology. See Science

Theory

The word "theory" is usually used precisely in physics. Theories are not conjectures but, rather, sets of equations whose solutions describe physical systems and their behavior. Using "theory" implies that there is some evidence for the validity. For example, a "theory of everything" would not only describe how things work, it would explain why things are the way they are. The name "theory of everything" is unfortunate in one sense: it does not tell how to deduce the behavior of complex systems from a knowledge of their components. In this book the name "final theory" has been used instead.

Transmit. See Mediate

Uncertainty Principle

The uncertainty principle is a consequence of quantum theory. It implies that a pair of observables cannot both be measured simultaneously to arbitrary accuracy. It can often be used to understand quantum theory results in a simple way.

Unification

Scientists have sought for centuries to unify the descriptions of apparently different phenomena by showing that they were due to the same underlying natural laws, and that complex levels of matter were made of simpler levels. This unification process is a subject of very active research for the forces of nature today. The possible unification of the strong, electromagnetic, and weak forces is called a "grand unification." There is a continuing effort to unify these forces with gravity. String theories seem to do that successfully.

Unstable Particle. See Decay

Vacuum

Any physical system will settle into the lowest energy state it can, which we call its vacuum state in particle physics. For most systems that is the state where the fields making up the system are zero, but theorists hypothesize that for systems containing Higgs fields the lowest energy occurs when the Higgs field takes on a constant value different from zero. The value of the Higgs field in that system is called its "vacuum expectation value." The universe will settle into the lowest energy state, which we can call "our string vacuum."

Vacuum Expectation Value. See Vacuum

Weak Charge. See Charge: Electric, Color, Weak

Weak Force

The weak force is described in Chapter 4. (*See also* Force.)

Wino

The supersymmetric partner of the W boson.

Zino

The supersymmetric partner of the Z boson.

Some Recommended Reading

Here are a few articles and books that may be of interest to readers of this book.

On the main subject, supersymmetry and its implications, there is little nontechnical good reading available. Some articles may be helpful: "Supergravity and the Unification of the Laws of Physics," by Daniel Z. Freedman and Peter van Nieuwenhuizen, *Scientific American,* February 1978; "Is Nature Supersymmetric?" by Howard E. Haber and Gordon L. Kane, *Scientific American,* June 1986; "Desperately Seeking SUSY," by Hans Christian Von Baeyer, *The Sciences,* September/October 1998; "The Dawn of Physics Beyond the Standard Model," by Gordon Kane, *Scientific American,* June 2003, p. 68, and Special Edition, January 2006, p. 5; and *"Mysteries of Mass,"* by Gordon Kane, *Scientific American,* July 2005, and Special Edition, January 2006. An earlier book is *Nature's Blueprint: Supersymmetry and the Search for a Unified Theory of Matter and Force,* by Dan Hooper (Smithsonian, 2008). In addition, a history of early supersymmetry work, with a mixture of technical and personal contributions, has been edited by myself and M. Shifman: *The Supersymmetric World—The Beginnings of the Theory* (World Scientific, 2000).

Readers who would like to learn more about the Standard Model of particle physics can turn to my earlier book, also for any curious reader: *The Particle Garden* (Helix Books, 1995). Some people with scientific training may want a somewhat more technical treatment of the Standard Model, which they can obtain from *Modern Elementary Particle Physics,* by Gordon L. Kane (Addison-Wesley, updated edition

1993), a book at the senior undergraduate or beginning graduate level. Two nice nontechnical studies of the history of the Standard Model, with considerable information about the experimental foundations, are *The Second Creation,* by Robert Crease and Charles Mann (Rutgers University Press, 1996), and *The Rise of the Standard Model,* edited by Lillian Hoddeson, Laurie Brown, Michael Riordan, and Max Dresden (Cambridge University Press, 1997).

Next I list some good books without detailed descriptions, alphabetically by author. Their subtitles provide information about what they cover.

A Zeptospace Odyssey: A Journey into the Physics of the LHC, by Gian Francesco Giudice (Oxford University Press, 2010).

The Elegant Universe: Superstrings, Hidden Dimensions, and the Quest for the Ultimate Theory, by Brian Greene (W.W. Norton & Company, 1999).

The Hidden Reality: Parallel Universes and the Deep Laws of the Cosmos, by Brian Greene (Vintage Books, 2011).

A Brief History of Time, by Stephen Hawking (updated tenth-anniversary edition, Bantam Books, 1998).

The Grand Design: New Answers to the Ultimate Questions of Life, by Stephen Hawking and Leonard Mlodinow (Bantam Press, 2010).

Perspectives on LHC Physics, edited by Gordon Kane and Aaron Pierce (World Scientific, 2008). This book has some nontechnical chapters, including an introduction to detectors by Maria Spiropulu.

Warped Passages: Unraveling the Mysteries of the Universe's Hidden Dimensions, by Lisa Randall (Harper Perennial, 2006).

Knocking on Heaven's Door: How Physics and Scientific Thinking Illuminate the Universe and the Modern World, by Lisa Randall (Harper-Collins, 2011).

Many Worlds in One: The Search for Other Universes, by Alex Vilenkin (Hill and Wang, 2006).

Dreams of a Final Theory: The Scientist's Search for the Ultimate Laws of Nature, by Steven Weinberg (Pantheon, 1992).

Index

Accelerators, 23, 82, 135–136
Acharya, Bobby, 150
AMS, 103
Annihilate, 31, 66, 100, 102, 103, 104
Anthropic questions, 136–142
Antielectrons, 31, 71
Antimatter, 31, 68
Antineutrinos, 71
Antiparticles, 19, 30–31, 40, 70, 71, 76
Antiprotons, 31, 103
Antiquarks, 103
Asymmetry, matter-antimatter, 40
ATLAS, 81, 112
Atomic size, 48
Atoms, 19, 20, 25, 26, 29, 96
    effective theory of, 48, 49
    size of, 54–55
Axions, 103–104

Backgrounds, 86, 89–90
Basic research, supporting, 134–136
Beyond the Standard Model, 38–43,
    63, 93, 119, 151
Big bang, 23, 40, 53, 66, 95, 118
Big questions, 17–18
Biology, 123–124
Black hole, 57

Bosonic dimensions, 74–75
Bosons, 19, 35, 38, 64–65, 108
    sparticles for, 73
    supersymmetry and, 61–62, 69, 72,
    74
Bottom quark (b quark), 32, 33, 112,
    115, 151
Broken supersymmetry, 75–79,
    124–125
Broken symmetry, 76

CERN, 9, 66–67, 81, 91, 135
Charge, 35–36
    conservation of, 23
    of Higgs field, 109–110
    particles and, 24
Charge conjugation invariance, 76
Charginos, 89, 126, 151
Charmed quark, 32, 33, 34
Chemical elements, 25
CMS, 81, 112
Cold dark matter, lightest superpartner
    and, 99–105
Colliders, 81, 82–85
    detection of lightest superpartner at,
    100–102
    future, 90

Colliders (continued)
   limits on experiments in, 90–94
   testing Standard Model and,
     66–67
Collider scale, 52
Colliding beam accelerators, 23
Compactification, 118, 119, 120–121,
    125–126
Compactified M/string theories, 94,
    119–121, 126–127
   Higgs boson and, 111
   Higgs boson mass and, 145–146,
    149–151
Compactified string theories, 108–109
Contact (Sagan), 54
Copernicus, 59
Cosmic rays, 83
Cosmological constant, 21, 141,
    142–145, 150
Cosmology, 46–47, 50, 51
Coulomb, Charles-Augustin de, 21
CP violation, 132, 134
Curie, Marie, 85

Dark matter, 17, 39, 47, 66, 84, 99–100,
    121
   cold, 99
   hot, 99
   LSP, 87
Darwin, Charles, 17, 131–132
Data, role of, 13–14, 36, 125–127
Decay, 22–24, 68
   γγ, 112
   of Higgs boson, 111, 112–113, 115,
    151
   of superpartners, 16–17, 86
Derivations, 12
Detectors, 82–85

   detection of cold dark matter,
    100–103
   funding for, 86
   recognizing superpartners, 86–88
   satellite, 102–103
Deuterium, 96
Dimensions
   extra, 118, 120–121, 125, 131,
    145–146
   fermionic, 74
   large extra, 131
   space-time, 75, 118, 119, 122, 149
   superspace, 72–75
Dirac, Paul, 30, 50, 71, 76
Direct detection, 101
Distance scales, 46–52, 51–52
   Standard Model and, 58–59
   See also Planck scale
Doppler shift, 96–97
Doubling of particles, 31, 146
Down quarks, 24, 34, 70, 95
"Dust," 96

Eames, Charles and Ray, 58
Effective theory(ies), 46
   of cosmology, 46–47, 50
   human scale, 59
   organizing by distance scales, 46–52
   Planck scale, 54–59
   supersymmetry and, 52–54
Einstein, Albert, 4, 6, 7, 28, 56, 70, 75,
    83, 118
Electrical fields, 28
   limits on, 92–93
Electrical force, 20, 21
Electric charge, 21, 23, 30, 31, 34,
    48–49, 50, 70, 76, 109, 110
Electromagnetic fields, 28, 36, 110

Electromagnetic forces, 21–22, 25, 47–48, 65, 67, 70
Electrons, 3, 16, 19, 23, 27, 29, 37, 66, 70, 95
 detectors and, 83
 discovery of, 20
 masses of, 12–13
 neutrinos and, 34, 71
 photons and, 21–22, 36
 properties of, 48
 quarks and, 24, 25–26
 in space, 103
Electroweak symmetry, 150
E=mc^2, 7, 84
Emergent properties, 50
Energy
 conservation of, 23
 of Higgs field, 110
 mass and, 56
 Planck, 57, 58
 in universe, 96, 97
Equations, 2–3, 7, 9, 11–14
 properties of, 12–14
 solutions and, 12–14, 144
Errors
 experimental, 15
 systematic, 15
Evolution, 123, 139
Experimental input in primary theory formulation, 69
Experiments
 errors in, 15
 limits on, 90–94
 relative to theory, 42–43
Explicit detection, 104

Families, 32–33, 41
Family problem, 33

Faraday, Michael, 21, 28
FERMI, 103
Fermi National Accelerator Laboratory (Fermilab), 112, 115
Fermionic dimensions, 74–75
Fermions, 19, 35, 38, 64–65, 108, 109
 supersymmetry and, 61–62, 69, 72, 74
Fields, 23–24, 27–30, 36
Final theory, 49, 52, 123, 140
 Planck scale and, 55–56, 58
 possibility of proving, 129–132, 146–147
 supersymmetry and, 61
 testing, 133–134
Forces, 4, 6–7, 20–22, 24–27, 36, 138–139
 unifying, 39, 65, 67
Friction, 47–48
Funding
 basic research, 134–135
 colliders/experiments, 90–91

Galaxies, 39, 47, 59, 96
Galileo, 2
Gauguin, Paul, 1, 2, 20
General relativity, 75
Generic, 119, 126
Geometrical symmetry, 73–74
Georgi, Howard, 113
γγ decay, 112
Gluinos, 89, 94, 126, 151
Gluons, 22, 24–25, 36, 95–96, 109
 mass of, 35
Gödel, Kurt, 130
Gravitational field, 28
Gravitational force, 20, 21, 47, 50, 141
 mass and, 23

Gravitational force (continued)
  Planck's scale and, 56, 57, 58
  on star, 96
  unification of forces with, 65, 67
Gravitinos, 72, 75, 94, 124
  mass of, 151
Gravitons, 22, 23, 75, 124
  mass of, 35
Gravity
  general relativity and, 75
  M/string theory and, 117, 122
  super-, 75
  supersymmetry and, 67
  theory of, 27–28
Ground state, 109, 121

Half-integer spin, 37–38
Havel, Vaclav, 137
Hawking, Stephen, 141
Helium, 96
Heterotic, 119
Hidden supersymmetry, 75–79,
    124–125
Hidden symmetry, 76
Hierarchy problem, 114–115
Higgs, Peter, 107
Higgs boson, 19, 23, 33–34, 77, 94,
    109, 110–114
  decay of, 111, 112–113, 115, 151
  discovery of, 3, 81, 84, 107–108
  hierarchy problem, 114–115
  mass of particles and, 35
  value of mass, 145, 149–151
Higgs field, 20, 39, 63–64, 77–78, 108,
    109–114
Higgsinos, 72, 89
Higgs interaction, 20
Higgs mass matrix, 150

Higgs mechanism, 77–79, 110–114,
    150
  mass from, 108, 115
  supersymmetry and, 89
Higgs physics, 35, 39, 107
  compactified string theories and,
    108–109
  Standard Model and, 63–64, 77,
    107–108, 111, 113–114
  supersymmetry and, 77–79
Hot dark matter, 99
How understanding, 50, 123–124
Human resources, for particle physics,
    92
Human scale, 47–48, 59
Hydrogen atoms, 96

Incompleteness theorem, 130
Indirect detection, 103
Inputs, effective theory and, 49–50
Integer spin, 37–38
Internal symmetry, 74
International Journal of Modern Physics
    (Kane, Acharya, and Kumar), 150
Investment in science, 91, 114, 136

Kepler, Johannes, 2
Kepler's laws, 39
Kumar, Piyush, 150

Lagrange, Joseph Louis, 29
Lagrangian, 28–29, 35, 143–144, 150,
    151
Lamb shift, 36
Large Hadron Collider (LHC), 9, 83,
    85, 118, 119
  detection of superpartners and, 17,
    81, 86, 89–90, 92, 126

Higgs boson observed at, 81, 111,
115, 149
LHCb, 151
lightest superpartners at, 100
LEP collider, 66–67
Leptons, 32–33, 41, 107, 108
decay of, 68
mass of, 35
sparticles for, 73
LHCb, 151
Light, 36
gravitational force and, 57
Lightest superpartner (LSP), 39–40, 66,
95
cold dark matter and, 99–105
as matter, 98–99
stability of, 87
Lithium, 96
Lu, Ran, 150
Lucretius, 27, 52

Magnetic field, 28
Magnetic force, 20, 21
Manifold, 120
Mass, 22–24, 23
energy and, 56
gravitino, 124
Higgs boson, 145, 149–151
from Higgs mechanism, 108, 115
of lightest superpartner, 99
of neutrinos, 40–41, 97–98
of particles, 12–13, 24, 35, 38–39,
79, 108
Planck, 41, 56, 64
production of particle and, 84, 85
of superpartners, 79, 86
Matter, 68
superpartner, 98–99

total, in universe, 96–97, 98
See also Cold dark matter; Dark
matter
Maxwell, James Clerk, 21, 65, 70
Measurement, 15
Meson, decay of, 151
Meters, 51–52
Minimal anthropic questions, 137–138
Missing energy effect, 87
Models, 7–8
Moduli, 119, 120, 126
Moduli bosons, 120
Moduli fields, 120
Moduli masses, 151
Molecules, 25
M/string phenomenology, 121
M/string theory, 53, 117–121
anthropic questions and, 139–140
axions and, 103
cosmological constant and, 143–144
LSP stability and, 87
overview, 122–124
role of data, 125–127
supersymmetry and, 67–68,
124–125
testing, 53, 132–133
types of, 119
See also Compactified M/string
theories
M/string vacuum, 121
"M/string Vacuum Project," 121
M-theory, 119
Multiverse, 144–145
Muons, 12–13, 32, 33, 83

National Institutes of Health, 136
National Science Foundation, 121
Nature, theory of, 4–8

Neutralinos, 89, 126, 151
Neutrinos, 19, 31–32, 32–33, 34, 66, 70, 71, 95, 97–98, 103
   mass of, 40–41
Neutrons, 19, 22, 24, 40, 96
Newton, Isaac, 4, 20, 27
Newtonian science, 9–10
Newton's constant, 55–56
Newton's laws, 6, 96
Non-Abelian gauge theories, 34
Nonminimal anthropic questions, 138–140
Nuclear force, 20, 22, 25
Nucleosynthesis, 118, 121
Nucleus(i), 19, 20, 25, 26, 29, 96
   effective theory of, 48, 49, 50

PAMELA, 103
Parity violation, 36, 121
Particle collider, 84
Particle physics, 42, 51
   goals of, 123
Particles, 4–8, 19, 24–27
   at beginning of universe, 40
   creation in collisions, 23
   decay and, 22–23
   detectors and colliders and, 82–85
   fields and, 27–30
   masses of, 12–13, 22, 24, 35, 38–39, 79, 108
   quanta and, 23–24
   superpartners, 16–17
   in universe, 97–100
   See also Superpartners (sparticles)
Pauli, Wolfgang, 31
Periodic table of elements, 48, 136
Perspectives in Higgs Physics, 113
Photinos, 72, 89, 100

Photons, 21–22, 22–23, 24, 27, 29, 37, 66, 95, 97, 109
   axions and, 104
   detection in space, 103
   detectors and, 83
   electrons and, 36
   mass of, 35
   number of, in universe, 97
Planck, Max, 51, 56–57
Planck energy, 57, 58
Planck length, 51–52, 56, 57
Planck mass, 41, 56, 64
Planck scale
   M/string theory and, 117, 120–121, 124, 125
   physics of, 54–59
   Standard Model and, 64–65
   supersymmetry and, 53, 68–69, 73
Planck's constant, 48, 49, 54, 55, 141
Planck time, 56, 57
Planck units, 56–57
Planets, 39, 40, 47, 66, 75, 96
Positrons, 31, 103
Postdiction, 15
Powers of Ten (film), 58
Predictions, 14–16, 34–36, 126
Primary theory, 52
Properties, emergent, 50
Protons, 19, 20, 22, 24–25, 40, 58, 66, 96
   decay of, 68

Quanta, 23–24, 28
   of Higgs field, 109, 110
Quantum fields, 120
Quantum fluctuations, 109
Quantum jump, 28
Quantum theory, 4–5, 6, 21, 28

electromagnetic interaction and, 30

Standard Model and, 35–36

Quarks, 3, 16, 19, 22, 23, 27, 29, 32–33, 34, 41, 95–96, 107, 108

bottom (b), 32, 33, 112, 115, 151

charmed, 32, 33, 34

decay of, 68

electrons and, 24, 25–26

Higgs physics and, 78

mass of, 35

sparticles for, 72, 73

strange, 32, 33

top, 32, 33, 34, 64, 78, 86

Radioactivity, 20

Range of variables, 9, 10

Rare decays, 68, 151

Reductionist, 46

Relativistic invariance, 6, 123, 133

Relativistic quantum theory, 6

Relic density, 104

Relics, 95

Research in progress, 8–11

Richter, Burton, 135

Rules, 4–8

Sagan, Carl, 54

Sakharov, Andrei, 40

Satellite detectors, 102–103

Science

boundaries of, 17–18

implications of final theory for, 146–147

open-ended except for particle physics and cosmology, 51

Scientific method, 1–2

Segmented approach, to study of physical world, 45–46

Selectrons, 72, 89

Sfermion, 72

Shakespeare, William, 146

Signature

of superpartners, 89–90

of supersymmetry, 86

SLC collider, 66–67

Sleptons, 89, 94

Smatter, 71

Smuons, 89

Sneutrino, 72

Solutions, 10, 12–14, 144

Space, 61–62

detection of cold dark matter in, 102–103

supersymmetry and, 67, 72–75

Space-time dimensions of string theory, 118, 120–121, 122

Sparticles. *See* Superpartners (sparticles)

Special relativity, 4–6, 7, 21, 30, 70, 118

Speed of light, 5, 28, 55, 57, 141

Spin, 19–20, 37–38, 61

superpartners and, 70, 71

Spinoffs, 135–136

Spin zero, Higgs boson and, 111, 112

Squarks, 72, 89, 94, 126

Standard Model, 7–8, 19–20, 26

distance scale of, 58–59

experimental foundations of, 36–37

extension of, 10, 38–43

forces and, 21

Higgs boson and, 33–34

Higgs physics and, 63–64, 77, 107–108, 111, 113–114

Lagrangian of, 29–31

predictions of, 34–36

scale of, 52, 64–65

Standard Model (continued)
superparticles and, 70–71
supersymmetric, 7, 53, 64, 65, 67,
72, 77–79, 108, 124, 125, 139
testing using colliders, 66–67
Stars, 39, 40, 47, 50, 59, 66, 96, 99,
140–141
Start-ups, 136
Strange quark, 32, 33
String theory, 10, 34. See also M/string
theory
Strong charge, 36
Strong force, 22, 24, 36, 67, 138–139
Subfields, 8–9, 10
Superconducting SuperCollider, 91,
115, 134
Supergravity, 75
Superpartners (sparticles), 16–17,
70–72, 73
decay of, 16–17, 66
experiments on, 84
masses of, 79, 86
personalities, backgrounds,
signatures of, 89–90
production/detection of, 84–85
recognizing, 86–88
superspace and, 74–75
terminology, 72
unifying forces and, 65, 67
See also Lightest superpartner (LSP)
Superspace, 72–75
Supersymmetric standard model, 7, 53,
64, 65, 67, 72, 77–79, 108, 124,
125, 139
Supersymmetry, 3, 10, 14, 18, 20, 35
dark matter and, 66
as effective theory, 52–54
evidence for, 69

extension of Standard Model and,
39, 40, 41
fermions and bosons and, 61–62, 72,
74
hidden or "broken," 75–79, 124–125
Higgs boson and, 33
Higgs mechanism and, 111
Higgs physics and, 77–79
impact of, 63–70
M/string theory and, 67–68
origin of idea, 61–62
Planck scale and, 68–69, 73
as space-time symmetry, 61–62,
72–75
superpartners and, 16–17
Supersymmetry theorists, 113–114,
121, 125
Swampland, 144
Symmetry, 34, 76
geometrical or space-time, 73–74
hidden or "broken," 14, 35, 76
internal, 74
Systematic errors, 15
System of units, 54

Tau, 12–13, 32, 33
Tau leptons, 112
Telescopes, 82
10/11D theory, 144
Testing
final theory, 133–134
M/string theory, 117–118, 122–123,
125–126, 132–133
predictions, 14–16
supersymmetry, 53
Tevatron collider, 112, 115
Theory, 6, 7, 8
discarding, 126–127

final, 49, 52, 55–56, 58, 123
primary, 52
proof and, 131
relative to experiment, 42–43
testing, 9, 10, 14–15
See also Effective theory(ies); Final
theory
Theory of Everything, 52
The Theory of Heat Radiation (Planck),
56–57
Time, 61–62
Planck, 56, 57
supersymmetry and, 67, 72–75
Top quarks, 32, 33, 34, 64, 78, 86
Top squark, 72
Type II, 119

Understanding, how *vs.* why, 50
Unification, 46, 64–65, 71, 75, 150
Units, system of, 54, 56–57
Universal constants, 54, 55
Universe
  age of, 40, 68
  anthropic questions and, 139,
    140–142
  cold dark matter of, 99–105
  composition of, 68
  creation of, 95–96, 141
  energy in, 96, 97
  expansion of, 95, 96–97, 142–143
  indifference of, 47
  limits to understanding of, 129–132,
    134–136
  multiverse, 144–145
  particles in, 97–100
  search for understanding of, 1–3,
    146
  size of, 68

total matter in, 96–97, 98
Up quarks, 24, 27, 32, 33, 34, 70, 95
Up squark, 72

Vacuum expectation value (vev), 109
Vacuum state, 121
Vafa, Cumran, 144

W bosons, 3, 22, 25, 36, 71, 78, 86, 89,
  100, 107, 108, 109, 111
  mass of, 35
Weak charges, 36
Weak force, 20, 36, 67
Weinberg, Steven, 52, 143
Why understanding, 50, 123–124
Winos, 72, 89, 94, 100
W mass, 79
World Wide Web, 135

Z bosons, 3, 22, 36, 86, 89, 100, 107,
  108, 109, 111
Zero spin, 37
Zheng, Bob, 150
Zinos, 89, 100
Z mass, 79